THE SCIENCE OF SUGAR CONFECTIONERY

RSC Paperbacks

RSC Paperbacks are a series of inexpensive texts suitable for teachers and students and give a clear, readable introduction to selected topics in chemistry. They should also appeal to the general chemist. For further information on available titles contact:

Sales and Customer Care Department, Royal Society of Chemistry,
Thomas Graham House, Science Park, Milton Road, Cambridge CB4 0WF, UK
Telephone: +44 (0) 1223 432360; Fax: +44 (0) 1223 423429; E-mail: sales@rsc.org

Recent Titles Available

The Chemistry of Fragrances
compiled by David Pybus and Charles Sell
Polymers and the Environment
by Gerald Scott
Brewing
by Ian S. Hornsey
The Chemistry of Fireworks
by Michael S. Russell
Water (Second Edition): A Matrix of Life
by Felix Franks
The Science of Chocolate
by Stephen T. Beckett
The Science of Sugar Confectionery
by W.P. Edwards

Future titles may be obtained immediately on publication by placing a standing order for RSC Paperbacks. Information on this is available from the address above.

RSC Paperbacks

THE SCIENCE OF
SUGAR CONFECTIONERY

W.P. EDWARDS

*Bardfield Consultants,
Essex, UK*

RS•C
ROYAL SOCIETY OF CHEMISTRY

ISBN 0-85404-593-7

A catalogue record for this book is available from the British Library

Published by The Royal Society of Chemistry,
Thomas Graham House, Science Park, Milton Road,
Cambridge CB4 0WF, UK

For further information see our web site at www.rsc.org

Typeset by Paston Prepress Ltd, Beccles, Suffolk
Printed by Athenaeum Press Ltd, Gateshead, Tyne and Wear, UK

Preface

While most people have eaten sugar confectionery at some time few people know the underlying science. Almost all sugar confectionery was developed not from an understanding of the science but by confectioners working by trial and error. In many cases this empirical knowledge was obtained before any scientific understanding was available.

There is one exception to this rule and that is where products have been made to resemble sugar confectionery but are free of sugars. This small area has absorbed more scientific effort than the rest of sugar confectionery put together.

This book is intended for everyone who has eaten sugar confectionery and wondered what the science behind it is. The work is not intended as a manual of methods for making confectionery but does give illustrative examples of manufacturing methods.

Some simple recipes have been included to allow readers to do some small scale confectionery making. Experiments can be based on these recipes to study a number of areas relating to confectionery making.

This book has to be dedicated to the largely anonymous confectioners who invented most sugar confectionery products. It is also dedicated to my old friend Brian Jackson.

Contents

Chapter 1

Introduction

The confectionery industry divides confectionery into three classes: chocolate confectionery, flour confectionery and sugar confectionery. Chocolate confectionery is obviously things made out of chocolate. Flour confectionery covers items made out of flour. Traditionally, and confusingly, this covers both long life products, such as biscuits, in addition to short-life bakery products. Sugar confectionery covers the rest of confectionery. In spite of the above definition, liquorice, which does contain flour, is considered to be sugar confectionery. The confectionery industry has created many confectionery products that are a mixture of categories, *e.g.* a flour or sugar confectionery centre that is covered with chocolate. There is another category that is sometimes referred to as 'sugar-free sugar confectionery'. This oxymoron refers to products that resemble sugar confectionery products but which are made without any sugars. The usual reason for making these products is to satisfy special dietary needs. A better name might be 'sugar confectionery analogues'.

The manufacture of confectionery is not a science-based industry. Confectionery products have traditionally been created by skilled craftsman confectioners working empirically, and scientific understanding of confectionery products has been acquired retroactively. Historically, sugar confectionery does have a link with one of the science-based industries – pharmaceuticals. In the eighteenth century, sugar confectionery products were made by pharmacists as pleasant products because the active pharmaceutical products were unpleasant. The two industries continue to share some technology, such as making sugar tablets and applying panned sugar coatings. There are products that although apparently confectionery are legally medicines. This usually applies to cough sweets and similar products. In the United Kingdom these products are regulated under the Medicines Act and require a product licence. This means that all the ingredients for the product are specified and cannot easily be altered. The dividing line between confectionery and medicines is not uniform in all countries.

One reason that confectionery making is not a science-based industry is the very long product life. The Rowntree's fruit pastille was invented in 1879 and was first marketed in 1881. This product is still one of the leading sugar confectionery lines in the UK today (1999), and it appears that it will continue to be sold into the 21st century. The man who invented it, Claud August Gaget, knew nothing of proteins or the peptide bond. In 1879 very little was known about proteins in scientific circles so there was no scientific basis from which to work.

FOOD LAW

Legislation affects all parts of the food industry. In Great Britain, modern food law developed from the Food and Drugs Acts. Such legislation came about after an outbreak of arsenic poisoning among beer drinkers, the cause of which turned out to be the glucose that had been used in making the beer – the glucose had been prepared by hydrolysing starch with sulfuric acid. The acid had been made by the lead chamber process from iron pyrites which contained arsenic as an impurity. The approach subsequently adopted was that all foods should be 'of the substance and quality demanded'. This was obviously intended to cover any future problems with other contamination, and not necessarily with arsenic. Other countries, particularly those whose legal systems follow Roman rather than Anglo-Saxon law, have tended to more prescriptive laws.

The British approach is to allow any ingredient that is not poisonous unless, of course, the ingredient is banned. Additives are regulated by a positive list approach: unless the substance is on the permitted list it cannot be used. There are anomalies where a substance can be legal in foods but which is not permitted to be described in a particular way. An example of this is the substance glycherrzin, which is naturally present in liquorice and has a sweet flavour. It would be illegal to describe it as a sweetener as it is not on the permitted sweetener list. Glycherrzin is permitted as a flavouring, however, and can be added to a food, which makes the overall product taste sweeter than it would without the addition. Conversely, the protein thaumatin is permitted as an intense sweetener yet, in practice, it has been found that thaumatin has more potential as a flavouring agent. It would have been much easier and cheaper to obtain approval for thaumatin as a flavouring rather than as a sweetener.

The British system does not automatically give approval to new ingredients merely because they are natural. This is in contrast with the position in some other countries – there will always be grey areas. One example is the position of the oligo-fructose polymers which are naturally present in chicory. Chicory is undoubtedly a traditional food ingredient; however, the oligo-fructoses extracted from it cannot necessarily be described as such. If the fructose polymers are hydrolysed to fructose then that is a permitted food ingredient. However, if they are partially hydrolysed then what is the status of the resulting product? The issue of fructose polymers is further complicated because one of the properties that is interesting is that they might not be completely metabolised. If that is the case then they would be considered as additives rather than ingredients. Additives need specific approval whereas ingredients do not.

Unlike chocolate confectionery, sugar confectionery is free of legal definitions. Terms such as 'pastille' or 'lozenge' although they have an understood meaning, at least to those in the trade, are sometimes applied to products that are not strictly within that understood meaning, *e.g.* there are products that are sold as pastilles but which are, in fact, boiled sweets. Butterscotch must contain butter, but gums do not have to contain any gum.

THE SCOPE OF SUGAR CONFECTIONERY

The confectionery industry is vast. It ranges from small shops, where the product is made on the premises, to branches of the largest companies in the food industry. Probably because sugar confectionery keeps well without refrigeration it has been a global market for many years. In spite of this there are distinct national and local tastes in sugar confectionery. A British jelly baby may resemble a German Gummi bear but the taste is quite different – curiously, the British jelly baby was invented by an Austrian confectioner. Similarly, the gum and gelatine pastilles made in France and Britain are very different, yet the leading British brand was invented by a French confectioner.

HEALTH AND SAFETY

Sugar confectionery is not an inherently dangerous product but several points should be made. Some sugar confectionery products are made at high temperatures, *e.g.* 150 °C, which is hotter than most forms of cookery even if it is not a high temperature by chemical standards. Precautions must also be taken to prevent contact between people and hot equipment or products. Sugar-containing syrups not only have a high boiling point but they are by nature sticky and a splash will tend to adhere. Precautions must be taken to prevent splashes and also to deal with any that occur. In the event of a splash, either plunging the afflicted area into cold water or holding it under cold running water is the best first aid. A sensible precaution is to make sure that either running water or a suitable container of water is always available.

Most sugar confectionery ingredients are not at high risk of bacterial contamination. However, some ingredients are prone to bacterial problems; examples are egg albumen and some of the gums and gelling agents. In handling these materials, precautions need to be taken so that they do not contaminate other ingredients or any finished product. Confectionery ingredients should be food grade and any confectionery being made to be eaten should be prepared using food grade equipment and not in a chemical laboratory. It must also be ensured that dusts from handling the ingredients do not cause eye or lung irritations. Some confectionery ingredients, although perfectly edible and of good food grade, can cause irritation if inhaled.

Chapter 2

Basic Science

There are several aspects of science which are fundamental to sugar confectionery. They are discussed here.

STABILITY

Sugar confectionery products keep well compared with most other food products. Their long life ensues because spoilage organisms cannot grow, and the reason that they cannot grow is because the moisture content is too low.

Water Activity

The relevant parameter is not only the water content but also the water activity. Water activity is a thermodynamic concept which accounts for the fact that materials containing different water contents do not behave in the same way, either chemically or biologically. It reflects the ability of the water to be used in chemical or biological reactions, and it is the concentration corrected for the differences in the ability of the water to undertake chemical reactions. If a non-volatile solute is dissolved in water then the vapour pressure decreases in a specific way for a perfect mixture. A thermodynamically ideal substance always has an activity of unity.

Originally, water activity could not be measured directly. One method was to measure the weight loss of a product held at a range of controlled relative humidities, which also has the effect of holding the product over a range of water activities. If a product is held at its own water activity it neither gains nor loses weight, and this point is described as its equilibrium water activity.

Equilibrium Relative Humidity (ERH)

The is term is normally abbreviated to ERH. The ERH can be deduced by extrapolating the weight loss data over a range of water activities for values greater and less than those actually measured for the product. Where the two lines intersect lies the water activity of the product. This extremely tedious and time-consuming method has largely been superseded by instruments that measure the water activity directly. The ERH still has practical importance since it is an indication of the conditions under which the product can be stored without deterioration.

Dew Point

A related property is the dew point which is the point at which condensation occurs upon cooling. When products are being cooled the temperature must not fall to the dew point otherwise condensation will occur on the product and product spoilage is likely.

COLLIGATIVE PROPERTIES

Boiling Points

Colligative properties are defined as those properties that depend upon the number of particles present rather than the nature of the particles. In sugar confectionery the most important of these is the elevation of boiling point. Because sugars are very soluble, very large boiling point elevations are produced, *e.g.* as large as 50 °C. Remembering that elevation of the boiling point is proportional to the concentration of the solute it is not surprising that the boiling point is used as a measure of the concentration and hence as a process control.

The boiling point of a liquid is the temperature at which the vapour pressure is equal to the atmospheric pressure. If the pressure is increased the boiling point will also increase whereas reducing the pressure will reduce the boiling point. Most sugar confectionery is made by boiling up a mixture of sugars in order to concentrate them. The use of vacuum here has several advantages. Energy consumption is reduced, browning is reduced and the whole process is speeded up. A common practice is to boil a mixture of sugars under atmospheric pressure to a given boiling point. A vacuum is then applied, which causes the mixture to boil under reduced pressure. This not only concentrates the mixture, but the latent heat of evaporation also cools the mixture rapidly, thus speeding up the production process since the product must ultimately be cooled to ambient temperature for further handling.

Another area where boiling points are important is with regard to steam. Most heating in a confectionery plant is done by saturated steam, *i.e.* steam at its boiling point. The temperature of steam can be regulated by controlling the pressure. One advantageous side effect of using vacuum boiling rather than boiling at atmospheric pressure is that lower steam pressures can be used because the boiling point has been reduced. These lower steam pressures produce a considerable saving in terms of the capital cost of steam boilers and pipework since they do not have to be built to withstand the higher pressures.

Measuring Vacuums

In controlling a process the level of the vacuum obtained controls the amount of water in the product. From a product stability consideration this is obviously important, and the level of vacuum applied can be measured in a number of ways. Although they may have been used in the past, mercury manometers, for obvious reasons, are no longer used. Nowadays, the commonest measuring instrument is probably the Bourdon gauge although various designs of pressure sensor are also available. Calibration of the gauge can be in a number of different units. It is common to find calibrations in units of length, *e.g.* inches or millimetres of mercury – this is a legacy of using a mercury manometer. Alternatively, units of pressure such as pounds per square inch (psi) or Newtons per square metre ($N\,m^{-2}$) are found. Another system is to use bars or millibars, where one bar is equal to one atmosphere.

pH

The pH scale is a convenient way of measuring acidity or alkalinity. The definition is

$$pH = -\log^{10}[H^+] \tag{2.1}$$

where $[H^+]$ is the concentration of hydrogen ions present in solution. This has the considerable advantage that it almost always gives a positive number. On the pH scale 14 is strongly alkaline whereas 1 is strongly acidic. The pH system does, however, imply that the solution is aqueous. When, as not infrequently happens in sugar confectionery, there is a higher concentration of sugar than water it does imply interesting questions regarding the result produced by a pH probe.

In sugar confectionery the pH of the product is important for a number of reasons. Fruit-flavoured products normally have some acid component added to complement the fruit flavour. Where a hydrocolloid is present the pH of the product can be critical otherwise the product will not be stable or it may not gel at all. If a hydrocolloid is held at its isoelectric point, *i.e.* the pH at which there is no net charge, then the hydrocolloid will likely come out of solution.

Buffers

Buffers are a convenient way of maintaining a fixed pH. Some natural materials, *e.g.* fruit juices, have a considerable buffer capacity. In confectionery, buffers are used as part of fruit flavour systems and when using high methoxyl pectin. In the case of high methoxyl pectin, gelling will only take place at high soluble solids and at acid pH. A buffer might consist of the sodium salt of a weak acid, *e.g.* boric acid, and the acid. Because the weak acid is only partially dissociated whereas the sodium salt is essentially completely dissociated, adding acid or alkali merely displaces the equilibrium of the weak acid solution, thus maintaining the pH. Almost all pHs can be obtained by appropriate choice of buffer.

POLARIMETRY

It is a property of any molecule possessing an asymmetric centre that when illuminated with plane-polarised light the plane of that light will be rotated – this is known as optical activity. Most sugar confectionery ingredients are optically active. As the amount of rotation is directly proportional to the amount of sugar present, measuring the optical rotation of a solution enables the concentration of sucrose or other sugars to be measured. When sucrose is broken into fructose and dextrose the rotation of polarised light is reversed; hence, this mixture of sugars is normally referred to as invert sugar. In confectionery factories, polarimeters such as in Figure 2.1 are used to check the concentration of products and components. This is a simple measurement to take and, although it is not a particularly accurate practice, it does suffice.

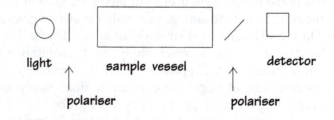

The angle between the polarisers is the optical rotation

Figure 2.1   *Measurement of optical rotation*

THE MAILLARD REACTION

Maillard reactions are responsible for the browning of sugars in the presence of amino acids. They are one of the key routes to flavour compounds in the whole of food science. In practice, any browning in foods is due to the Maillard reaction except where it is enzymic, *e.g.* the browning of a cut apple is enzymic and hence not a Maillard reaction.

The Maillard reaction is not a name reaction where all the details can be found in a text book: the term covers a whole range of reactions that occur in systems ranging from food to the life sciences. In sugar confectionery the problems with Maillard reactions are in preventing them where they are not wanted, *e.g.* in boiled sweets, and encouraging them where they are, *e.g.* in toffees. The name of the reaction goes back to Louis Camille Maillard who heated amino acids in a solution with high levels of glucose.*

Without doubt, the chemistry of the Maillard reaction is complex. It is complex not only because the reaction can give complex products but also because the starting materials are themselves complex. Most model systems involve studies of one reducing sugar being heated with one amino acid (Figure 2.2). A typical confectionery system, such as for a toffee, involves heating a mixture of proteins, usually from milk, with a mixture of reducing sugars and fats. In sugar confectionery manufacture the conditions of the reaction are likely to be high temperature but low water activity. In the early stages of the reaction, the free amino group of an amino acid, usually in a protein, condenses with the carbonyl group of a reducing sugar. The resulting Schiff bases rearrange by Amadori (Figure 2.3) or Heyns (Figure 2.4) rearrangements, the products being an *N*-substituted glycosylamine (if the sugar is, for example, glucose) and an *N*-substituted fructosylamine (if the sugar is a ketose such as fructose), respectively. In the advanced stages of the reaction, the rearrangement products degrade by one of three possible routes. They break down either *via* deoxysones, fission or Strecker degradation (Figures 2.5 and 2.6). The 1-deoxyglycosones and 3-deoxyglycosones can form reactive α-dicarbonyl compounds such as pyruvaldehyde and diacetyl by retro-aldolisation reactions. These reactive intermediates are then available to react with ammonia and hydrogen sulfide.

In the final stages of the reaction, brown nitrogenous polymers and copolymers form. The chemical nature of the compounds concerned is little known. It has been shown[1] that heating proteins and carbonyl-

---

*L.C. Maillard and M.A. Gautier, 'Action des acides amines sur les sucres: formation des melanoidines par voie methodique. *CR Acad. Sci.*, 1912, **154**, 66–68.

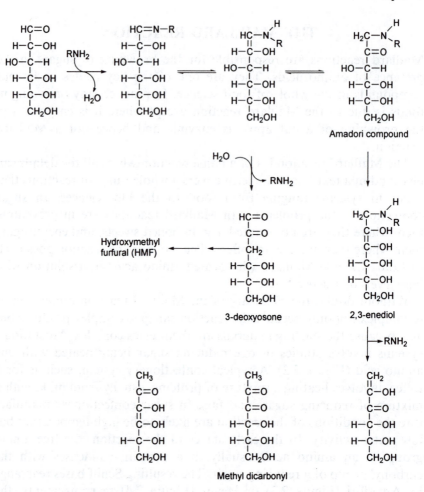

Amadori compound

3-deoxyosone

2,3-enediol

Hydroxymethyl furfural (HMF)

Methyl dicarbonyl

Figure 2.2 *The Maillard reaction*

containing compounds together causes protein gels to form. It is believed that these proteins become covalently linked to one another, and this sort of process could easily occur in toffee making. There are claims that the effect of the Maillard reaction is to reduce the availability of amino acids.

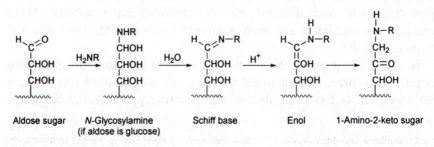

Aldose sugar   N-Glycosylamine   Schiff base   Enol   1-Amino-2-keto sugar
                    (if aldose is glucose)

Figure 2.3 *Amadori rearrangement*

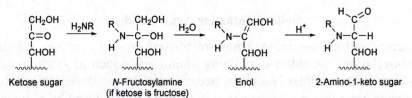

Figure 2.4 *Heyns rearrangement*

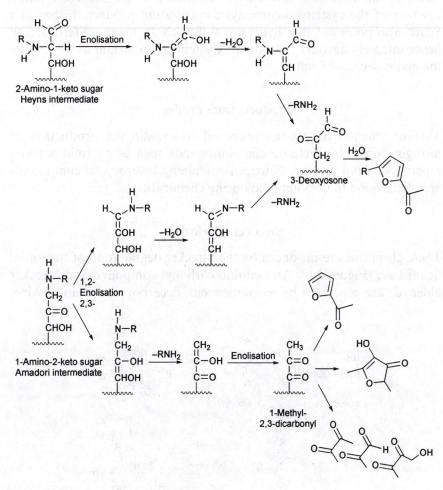

Figure 2.5 *Formation of furans and dicarbonyls*

As confectionery is only a minor part of the diet this is only a minor problem. If amino acids have undergone complicated reactions it is not too surprising that they are not biologically available in the finished product.

Sulfur-containing Amino Acids

Whereas sulfur-free amino acids are broken down to amines *via* decarboxylation, the sulfur-containing amino acids such as cysteine can undergo more complex reactions. Because cysteine produces a powerful reducing aminoketone, hydrogen sulfide could be produced by reducing mercaptoacetaldehyde or cysteine.[2] Alternatively, hydrogen sulfide could be produced alongside ammonia and acetaldehyde by the breakdown of the mercaptoimino-enol intermediate of the decarboxylation reaction of the cysteine-dicarbonyl condensation product. Fisher and Scott[2] also point out that hydrogen sulfide forms many odiferous, and hence intensely flavoured, products. Cysteine is important as it is one of the major sources of sulfur.

Products from Proline

Various schemes have been proposed to explain the production of nitrogen-containing heterocyclic compounds such as pyrrolidines and piperidines from proline. Nitrogen-containing heterocyclic compounds are often found to be potent flavouring chemicals.

Strecker Aldehydes

These chemicals are produced by the Strecker degradation of the initial Schiff base (Figure 2.6). An α-amino carbonyl compound and Strecker aldehyde are generated by rearrangement, decarboxylation and hydro-

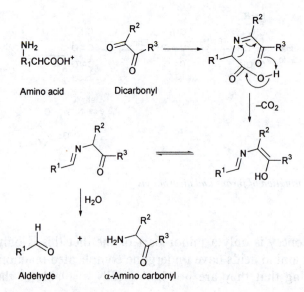

Figure 2.6   *Strecker degradation*

lysis. Thus the Strecker degradation is the oxidative deamination and decarboxylation of an α-amino acid in the presence of a dicarbonyl compound. Thus an aldehyde with one fewer carbon atom than the original amino acid is produced. The other class of product is an α-aminoketone. These are important as they are intermediates in the formation of heterocyclic compounds such as pyrazines, oxazoles and thiazoles. These heterocyclic compounds are important in flavours.

DENSIMETRY

The density of sugar syrups is used as a method of measuring the quantity of sugar present. It is possible to make very accurate measurements of density and for this confectioners often use simple hygrometers. The data obtained give very accurate information relating density to sugar concentration.

Some non-SI units are in use in this area. Rather than report a density, the ratio of the density of the syrup to that of water is used, *i.e.* the specific gravity. This of course makes the specific gravity a ratio and is hence without units. The percentage of sucrose by weight is sometimes reported in degrees Brix. The difference between reporting sucrose concentrations as weight/weight (w/w) and weight/volume (w/v) can be considerable. As an example, 50 g of sugar in 50 g of water is 50% sugar w/w, *i.e.* 50 Brix, but 50 g of sugar dissolved in water and made up to 100 ml is 50% w/v which is approximately 42% w/w. 50 g of sugar dissolved in 100 ml of water approximates to 33.3% w/w.

The Baume scale is still used in the industry where

$$\text{Baume} = M - (M/S) \tag{2.2}$$

for which M is a modulus and S is the specific gravity. In the UK, $M = 144.3$ whereas in the USA and parts of Europe $M = 145$.

Tables have been published relating Baume, Brix and specific gravity. As density is temperature-dependent it is necessary to bring the syrup to a fixed temperature. In practice it is more common to use temperature correction factors or tables. The relationship between density and concentration is slightly different for invert sugar or glucose syrups. The Brix scale is sometimes applied to products which are not sucrose syrups, such as concentrated fruit juice. Recipes are certainly in use which state 'boil to x Brix'. In practice, what these instructions mean is that the material should give the same reading as a sugar syrup of that concentration. As often happens in confectionery these practices have been proved to work empirically.

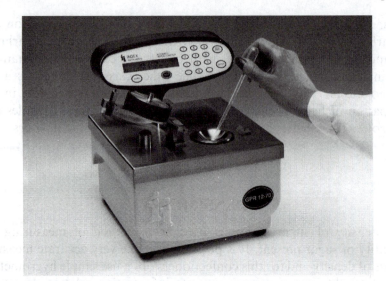

Figure 2.7 *Refractometer*

REFRACTIVE INDEX

Another commonly used control measure is that of refractive index. The refractive index of a substance is the ratio of the velocity of light in a vacuum to the velocity of light in the substance and is measured using a refractometer (Figure 2.7). When light passes from one medium to another the beam is refracted to an extent determined by the change in the refractive indexes of the two substances. The variation in refractive index with concentration for sucrose is well known. Similar but not identical variations occur for glucose and invert sugar syrups. In practice, refractometers calibrated to measure sucrose concentration are normally used regardless of the actual sugars present. Apart from the boiling point, the refractive index is the commonest control measure used in the manufacture of sugar confectionery. A refractometer is normally more expensive than a thermometer.

ANALYTICAL CHEMISTRY

The analytical chemistry that is applied to confectionery, as in other products, has changed enormously. High powered analytical techniques are now readily available.

WATER CONTENT

As already mentioned, the amount of water present is fundamental to the stability of confectionery products. Not surprisingly, the measuring of water contents is an important exercise and various methods are used. Some moisture content determinations using the oven drying technique are still carried out, although this sort of work is difficult since moisture contents are normally low and the samples can only be dried with difficulty. In particular, the problems are in drying the product within a reasonable time without charring it. Various other methods of water content determination are in use – one is the Karl Fischer titration.

In this system, a reagent prepared by reacting sulfur dioxide with iodine dissolved in pyridine and methanol is used:

$$3C_5H_5N + I_2 + SO_2 + H_2O \longrightarrow 2C_5H_5N^+I^- + C_5H_5N^+ SO_2O^-$$

$$C_5H_5N^+ SO_2O^- + CH_3OH \longrightarrow C_5H_5N \underset{H}{\overset{OSO_2OCH_3}{\diagup}}$$

Initially, the sulfur dioxide is oxidised by the iodine. This can take place only in the presence of an oxygenated molecule. The product can be regarded as pyridine–sulphur trioxide complex. In the next stage the methyl ester is formed. Thus, one molecule of water is equivalent to one molecule of iodine. The original Karl Fischer reagent was prepared with an excess of methanol where the methanol acted both as a solvent and as a reagent in forming the complex. This type of reagent tends to be unstable and so alternative forms of the Karl Fischer reagent have been developed where the methanol is substituted with ethylene glycol mono-methyl ether (methyl cellosolve). Versions of the reagent without pyridine are also available but the pyridine-free version tends to be less successful than the original form.

Although it is just about possible to perform Karl Fischer titrations in a fume cupboard using simple titration apparatus and the iodine colour as an indicator, special titration apparatus is normally used. The end point is normally measured electrically by applying a small voltage across two platinum electrodes. With this special apparatus, the sample is titrated with Karl Fischer reagent until the end point is reached when free iodine appears causing an increase in conductivity. This is detected electronically. The most modern Karl Fischer titrators aim for a high degree of automation. Some instruments have a macerator blade in the titration vessel to break up the sample. This is effective with brittle samples, such as boiled sweets, where the sample shatters on impact with the blade. It tends to be unsatisfactory with gum and jelly sweets: these tend to be rubbery and instead of shattering remain intact and only release their water content slowly.

Instrumental Methods

Water determinations tend to work well on instrumental analysis probably because water is radically different from other substances. Methods such as NMR and near infra-red are both applied to confectionery products.

NMR

Proton NMR is obviously likely to give an enormous range of signals from a typical confectionery product. For the analysis of water in confectionery, the NMR instrument used must be of low resolution whether it is of the original continuous form or of the later pulsed type. The aim of the exercise is to discriminate between the protons in water and those in other molecules. Fortunately, this is not too difficult.

Near Infra-red

Near infra-red spectroscopy (NIRS) uses that part of the electromagnetic spectrum between the visible and the infra-red regions. This area has the advantage that the instrumentation is nearest to visible instrumentation. The signals in the near infra-red come not from the fundamental vibrations of molecules but from overtones. Together with the instrumentational advantages, the occurrence of overtone signals means that the selection rules are relaxed and all possible absorbances occur. In general, NIRS measures overtones of stretches using OH for water and NH for protein. As water gives a response different from other substances this determination often works well.

Problems with Moisture Determination

It might be expected that measuring the moisture content of sweets being dried is easy. This is obviously a useful control measure in a factory where gums or pastilles are being stoved (see also Chapter 10). The problem with this measurement is that the sweet is not homogeneous. It is entirely possible to have a dried sweet where the outside has a solids content of 92% but the middle with a solids content of 86%. Any technique that is surface-biased can produce any value between 92 and 86% on a cross-section of the same sweet.

SUGAR ANALYSIS

An old-fashioned chemist would perform sugar analyses by Fehlings titration before and after inversion and polarimetry. A technique very valuable to analytical chemistry is gas chromatography (GC). However, as sugars are non-volatile it is not possible to use gas chromatography to analyse them directly. If sugars are to be analysed using a GC method they must first be derivatised to produce volatile derivatives.

The first big improvement in direct sugar analysis was the use of HPLC. The columns used were silica-based amino-bonded phases with a mobile phase of acetonitrile and water. Polymer-based metal-loaded cation exchange columns were also used. These methods worked well for small sugars but were hampered in several ways. The detector of choice used to be the refractive index detector since sugars do not have a UV absorption, except at short wavelength. (Short wavelength UV detection requires specially pure acetonitrile, and even then there are many interferences.) The refractive index detector precludes the use of gradient elution (which would enormously increase the separation power of any HPLC system). HPLC was not generally able to analyse a whole range of large and small saccharides in one chromatogram. In analysing the types of mixtures of sugar and glucose syrup common in sugar confectionery one problem was that the high molecular weight component of the glucose syrup was not eluted and periodically had to be washed from the column.

Ion chromatography has since become available and is now commonly used for sugar analysis. In this system a high performance anion exchange column is used at high pH. This separation works because neutral saccharides behave as weak acids. Table 2.1 shows some pK_a values.

Very conveniently, the technique can also handle the sugar alcohols such as sorbitol (see also Table 2.1). Detection is by pulsed amperometric detection, and the system works by detecting the electrical current generated by oxidation of the carbohydrate at a gold electrode. However, the oxidation products poison the surface of the electrode necessitating cleaning between measurements. The cleaning is carried out by raising the potential to oxidise the gold surface. This causes the oxidation product to desorb. Next the potential is lowered which reduces the electrode back to gold. Thus, the sequence of pulsed amperometric detection is measuring the current at the first potential and then applying

Table 2.1 pK_a *values of some common saccharides*

Sugar	pK_a
Fructose	12.03
Mannose	12.08
Xylose	12.15
Dextrose	12.28
Galactose	12.39
Sorbitol[a]	13.6

[a]A sugar alcohol.

a more positive potential to oxidise and clean the electrode, followed by another potential to reduce the electrode back to gold ready for the next detection cycle. In operation the three potentials are applied for a fixed duration, where there is also a charging current when changing potentials. The oxidation current is distinguished from the charging current by measuring it after the charging current has decayed. Integration of the cell current over time is used to obtain the carbohydrate oxidation current, and as the integration of current over time gives charge the value obtained is in Coulombs.

An important question is how this system is able to work with sugar alcohols and non-reducing sugars. The oxidation is catalysed by the electrode surface which means that the response is dependent upon the electrode potential of the catalytic state rather than the redox potential.

As the mobile phase in this system is normally a sodium hydroxide solution there is no need to handle or dispose of organic solvents. This is a particular bonus to some smaller sites which are not set up to use organic solvents.

An important issue for any laboratory analysing products rather than raw materials is extracting the material of interest from the product under analysis. Some sugar confectionery, such as boiled sweets, can be dissolved directly with little preparation whereas other materials, like toffees, require considerable extraction and clean up. If a toffee is to be analysed for sugars then the sugars have to be separated from the proteins and fat present. This is made more difficult by the fact that the system is by design a stable emulsion (see below). Most methods of sugar analysis require a clean aqueous extract to work on. One problem of working with HPLC columns is that minor components can accumulate on the column thus deactivating it. Ion chromatography has the advantage that it is relatively easy to clean the column as sodium hydroxide at high pH can be used.

As an example, the stages for analysis of a butterscotch are as follows. Dilute the sample 1:2000 with water and pass the solution through a 0.2 μm filter. The dextrose, fructose, maltose and maltotriose can then be measured directly using ion chromatography. In contrast, the same analysis by HPLC would probably require two chromatograms performed with different mobile phases using the amino column HPLC method.

EMULSIONS

An emulsion is a dispersed system of two immiscible phases. They are present in a number of food systems. In general, the disperse phase in an emulsion is normally in globules with diameters of between 0.1 and 10 μm. It is common to class emulsions as either oil in water (O/W) or water in oil (W/O). In sugar confectionery the emulsions usually encountered are oil in water, or perhaps more accurately oil in sugar syrup.

One of the most important properties of an emulsion is its stability. Emulsions normally break by one of three different processes: creaming (or sedimentation), flocculation or droplet coalescence. Creaming and sedimentation have their origin in density differences between the two phases, and emulsions often break by a mixture of the three main processes. The time it takes for an emulsion to break can vary from seconds to years.

Emulsions are not normally inherently stable since they are not a thermodynamic state of matter – a stable emulsion normally needs some material to give it its stability. Food law complicates this issue since various substances are listed as emulsifiers and stabilisers. Unfortunately, some natural substances that are extremely effective as emulsifiers in practice are not emulsifiers in law. An examination of those materials that do stabilise emulsions allows them to be classified as follows:

(1) surfactants;
(2) natural products;
(3) finely divided solids.

Some substances fall into more than one category. In a practical emulsion system the emulsifier should facilitate making the emulsion as well as stabilising it after formation. Some properties have opposite effects in the two areas. A high viscosity makes it harder to form an emulsion but obviously tends to stabilise the emulsion once formed.

THE CHEMISTRY OF OILS AND FATS

Fats are chemically triglycerides and can be regarded as the esters produced by the reaction of fatty acids with the trihydric alcohol glycerol. In practice, oils and fats are the product of biosynthesis. Some sugar confectionery contains oils or fats whereas other products, *e.g.* boiled sweets, are essentially fat-free. The traditional fat used in sugar confectionery is milk fat, either in the form of butter, cream, whole milk powder or condensed milk. Milk fat can only be altered by fractionating it, and while this is perfectly possible technically, there must be sufficient commercial and technical benefits to make it worthwhile. One problem with fractionation operations is that both the desirable and the undesirable fractions have to be used.

Whereas vegetable fats were used originally as a cheaper substitute for milk fat, the ability to specify the properties of vegetable fat has considerable advantages. This ability arises because of the science and technology available to the fat-processing industry. Some vegetable fats used in sugar confectionery are not tailor-made but are simply a vegetable fat of known origin and treatment. The commonest example is hydrogenated palm kernel oil (HPKO) which is often used in toffees.

Some fats go into confectionery as a component of other ingredients. The common example is nuts, which contain fats often of types such as lauric acid in addition to unsaturated fats. These fats are sometimes the origin of spoilage problems (see also page 22).

Classifications of Fatty Acids

Fatty acids consist of a hydrocarbon chain with a carboxylic acid at one end. They can be classified on the basis of the length of the hydrocarbon chain (Table 2.2) and whether there are any double bonds. Trivial names of fatty acids such as butyric, lauric, oleic and palmitic acids are in common use in the food industry. A form of short-hand is used to refer to triglycerides where POS is palmitic, oleic, stearic. If the chain length is the same an unsaturated fat will always have a lower melting point. Another classification of fats that is used is in terms of the degree of unsaturation of the fatty acids. Saturated fats are fats without any double bonds. Many animal fats are saturated, but some vegetable fats, *e.g.* coconut oil, are saturated also. Mono-unsaturated fats include oils like olive oil but also some partially hydrogenated fats. Polyunsaturated fats have many double bonds and include sunflower oil. Because they are

Table 2.2 *Trivial names for fatty acids*

Trivial name of fatty acid	Nomenclature
Butyric	$C_{4:0}$
Caproic	$C_{6:0}$
Caprylic	$C_{8:0}$
Capric	$C_{10:0}$
Lauric	$C_{12:0}$
Myristic	$C_{14:0}$
Palmitic	$C_{16:0}$
Stearic	$C_{18:0}$
Oleic	$C_{18:1}$
Linoleic	$C_{18:2}$
Linolenic	$C_{18:3}$
Arachidic	$C_{20:0}$
Gadoleic	$C_{20:1}$
Behenic	$C_{22:0}$
Erucic	$C_{22:1}$
Lignoceric	$C_{24:0}$

too unstable, polyunsaturated fats are not normally used in significant quantity in confectionery. A rare exception to this would be if a polyunsaturated oil, *e.g.* sunflower oil, was used for marketing reasons. Other occasions have occurred when sunflower oil has been used by those unaware of its chemistry. Traces of polyunsaturated oils do go into sugar confectionery as components of ingredients, *e.g.* nuts.

The Hydrogenation of Fats and Oils

In order to provide the right properties it is often necessary to reduce the degree of unsaturation of a particular fat or oil. This is achieved by hydrogenating the oil over a catalyst, usually nickel. The hydrogenation can be complete, which yields a saturated fat, or partial, which yields a partially hydrogenated or hardened fat. Partial hydrogenation tends to produce fats with *trans*-double bonds. A double bond is physically flat and does not permit rotation and thus the chemical groups are fixed in their relation (Figure 2.8). A *cis*-double bond is one where the hydrogen atoms are both on the same side. Similarly, a *trans*-double bond has them on opposite sides. Most naturally-occurring oils and fats have *cis*-double bonds; however, some *trans*-double bonds are found in milk fat and certain marine oils.

Fat Specifications

Apart from specifications as to origin, *e.g.* palm kernel oil, fats are normally supplied on the basis of established parameters. One of these is the iodine value. This reflects the tendency of iodine to react with double bonds. Thus the higher the iodine value the more saturated the fat is. An iodine value of 86 approximates to one double bond per chain, whereas an iodine value of 172 approximates to two double bonds per chain. Another parameter is the peroxide value. This attempts to measure the suscept-ibility of the fat or oil to free radical oxidation. The test is applied on a freshly produced oil and measures the hydroperoxides present. These

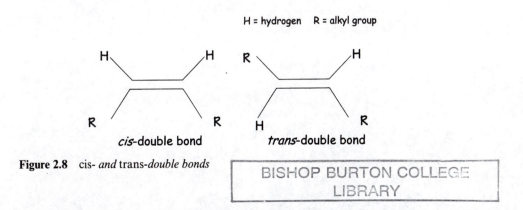

Figure 2.8   cis- *and* trans-*double bonds*

hydroperoxides are the first stage of the oxidation process. Obviously this test would not give reliable results if applied to a stale sample.

Deterioration of Fats

There are two processes that cause fats to deteriorate. One is normally a chemical process (oxidative rancidity), the other normally enzymatic (lipolytic rancidity).

In oxidative rancidity, oxygen, normally in the form of a free radical, adds across double bonds. This is a zero activation energy process so it is not inhibited by reducing the temperature. The end products of this process can be unpleasant in both taste and smell. Oxidative rancidity tends to appear suddenly and then progress rapidly, and may be prevented by using saturated fats.

Lipolytic rancidity is normally enzymatic, the enzymes responsible usually coming from bacteria or moulds. The effect of lipolytic rancidity is that the level of free fatty acid rises. The effect of this on the product depends very much upon the nature of the free fatty acid liberated. Low levels of free butyric acid from milk fat tend to enhance a toffee by giving it a more buttery flavour, whereas lipolysis of a lauric fat such as HPKO gives free lauric acid, which is an ingredient of, and tastes of, soap. This effect is very unpleasant.

REFERENCES

1. S. Hill and A.M. Easa, in *The Maillard Reaction in Foods and Medicine*, Royal Society of Chemistry, Cambridge, 1998, pp. 113–138
2. C. Fisher and T.R. Scott, *Food Flavours: Biology and Chemistry*, Royal Society of Chemistry, Cambridge, 1997.

Chapter 3

Ingredients

SUGARS

Sugar confectionery has developed around the properties of one ingredient – sucrose. Sucrose is a little unusual as a sugar as it is a non-reducing disaccharide. Its constituent monosaccharides are dextrose (or glucose – see page 26) and fructose, both of which are reducing sugars. One of the crucial properties of sucrose is that its solubility at room temperature is limited to 66%. This feature means that a sucrose solution is not stable against bacteria or moulds. As an asymmetric molecule sucrose rotates the plane of polarised light, and it is easily observable that if sucrose is heated with acid or alkali, or treated with the enzyme invertase, the optical rotation alters to the opposite direction. In fact, the rate of reaction can be measured by monitoring the optical rotation. This change in optical rotation is called inversion and occurs because the sucrose splits into fructose and dextrose. In practice, a small degree of inversion of sucrose normally occurs when sucrose is boiled up in water.

Sucrose is extracted either from sugar beet or sugar cane. Normally, the two sources are equivalent even though the trace impurities are different. There is one area where the two sources are not equivalent and that is regarding brown sugars. Cane sugar that has not been completely purified has a pleasant taste and can be used as an ingredient. Beet sugar, however, is not acceptable unless it is completely white. In some products, brown sugars or even molasses (the material left after sugar refining – see below) are used to add colour and flavour. Alternatively, in some products a less than completely white product is used simply to save money. Beet sugar refiners do produce brown sugars which are produced by adding cane sugar molasses to refined beet sugar, some specifications of which are given in Table 3.1. Brown sugars used in confectionery are carefully controlled products: they are not refined to a high degree of purity but they are produced with carefully controlled levels of impurity. Raw sugar is not normally used in confectionery, although there is one exception where very small tonnages of health food

Table 3.1 *Specifications for brown sugars*

	Light	Dark	Demerara
Appearance	Fine, light golden brown	Fine, dark golden brown	Coarse, golden brown
Typical solution colour (ICUMSA units)	6000	22 000	4000
Particle size (MA μm, typical)[a]	300	300	700
Reducing sugars (%, minimum)	0.3	0.3	N/A
Loss on drying (%, maximum)	0.7	2.8	0.5
Total sugars (%, typical)	99	96.4	99.4
Molasses addition (%)	2	8	1.3

[a] MA = mean aperture of measuring sieve.

confectionery are made using raw sugar. Presumably the customers for this class of product believe that some benefit is conferred by using the material in its raw state.

Confectionery factories normally use sucrose in a number of forms (for examples of particle size and forms commonly used in different confectionery products see Tables 3.2 and 3.3): granulated, *i.e.* crystalline milled sugar, icing sugar and possibly a 66% sugar syrup (see also

Table 3.2 *Particle sizes for different grades of sugar*[a]

Type of sugar	Biggest size	Smallest size
Granulated	400 μm	600 μm
Icing (milled sugar)	10–15 μm mean	10–15 μm mean
Coarse sugar	>55% above 1.18 mm	<5% below 850 μm
Medium sugar	<8% above 1.18 mm	<12% below 600 μm
Powdered sugar	17% max above 212 μm	23–55% below 53 μm
Ultra fine sugar	<45% above 355 μm	20–45% below 150 μm
Caster sugar	<10% above 425 μm	<22% below 212 μm
Non pareil sugar	<5% above 850 μm	<10% below 600 μm
Fine sugar	<7% above 850 μm	<13% below 425 μm

[a] Source: British Sugar.

Table 3.3 *Forms of sugar commonly used in sugar confectionery*

	White granulated	Speciality white granulated	Screened specialities	Milled specialities	Brown sugars	Liquid sugars	Syrups and treacles
Boiled sweets	yes	yes	yes	yes	no	yes	no
Toffees/fudges	yes	yes	yes	yes	yes	yes	yes
Gums/pastilles	yes	yes	yes	no	no	yes	no
Chewing gum	yes	no	no	yes	no	no	no
Liquorice	yes	yes	no	no	yes	no	yes

Table 3.4 *Properties of commercial liquid sugars**

Product	Colour (ICUMSA units)	Solids (%)	Reducing sugars (%)
Liquid sugar no. 25	25 max.	67.0 ± 0.5	0.05 max. on solids
Partial invert syrup I	100 max.	66.0 ± 1.0	55.0 ± 1.0
Partial invert syrup II	100 max.	77.0 ± 1.0	44.5 ± 0.5
Medium invert syrup	100 max.	74.0 ± 0.5	36.5 ± 1.0
Invert syrup I	80 max.	80.3 ± 0.3	77.0 ± 1.0
Invert syrup II	80 max.	78.0 ± 0.3	74 min.
Invert syrup III	100 max.	68.0 ± 1.0	65 min.
Invert syrup IV	100 max.	75.9–76.9	74 min.

*Source: British Sugar.

Table 3.4). Sugar is normally supplied to the factory in the granulated form. Sugar syrup is not stable and the economics of transporting large weights of water are not favourable. Powdered, milled sugars have the problem that they are potentially an explosive dust and must be handled with appropriate precautions. Some factories mill their own sugar on site whereas others have the sugar supplied pre-milled.

Molasses and Treacle

Molasses is the material left when no more sugar can be extracted from the sugar beet or cane. Beet sugar molasses has an unpleasant taste and is not normally used for human food. Cane sugar molasses does have some food use, normally in the form of treacle which is clarified molasses. The ratio of sugar to invert sugar in treacle can be altered to some extent to assist with product formulation. In practice, different sugar syrups are blended with the molasses to give the desired product. Treacle is normally stored at 50 °C to maintain liquidity.

Invert Sugar

Invert sugar is only encountered as a syrup. The fructose in the mixture will not crystallise, so attempts to crystallise invert sugar yield dextrose. Invert sugar overcomes one of the big drawbacks of sucrose in that invert solutions can be made at concentrations as high as 80%. These solutions have a sufficiently low water activity that they do not have biological stability problems. More importantly, invert sugar can be mixed with sucrose and concentrated sufficiently to yield products that not only have a sufficiently low water activity to be stable but that also will not crystallise. Adding invert sugar to a formulation lowers the water activity but makes the product hygroscopic (prone to absorbing water from the surrounding atmosphere). Some old-fashioned sugar confectionery products do not contain invert sugar as an ingredient but rely on the effect of heating sucrose in the presence of acid to generate some invert sugar *in situ*. The use of invert sugar has declined since glucose syrup is cheaper and for some uses has superior properties, although some confectioners take the view that invert syrup improves the flavour of certain products. There is, however, another reason that encourages the use of invert sugar. Sugar-containing wastes can often be treated to produce invert sugar syrup. If a sugar solution is poured down a factory drain this generates a substantial charge for treating the resulting effluent. At the time of writing a tonne of sugar costs around £400. Allowing this tonne of sugar to become waste generates a further cost of £200. If the sugar can be recovered to produce invert, not only is the invert available as an ingredient, replacing some purchased material, but the £200 per tonne disposal cost is also avoided.

Glucose Syrup (Corn Syrup)

The ingredient known in the UK as glucose syrup has largely replaced invert sugar as a confectionery ingredient. Indeed, some sugar confectionery products contain more glucose syrup solids than sucrose. In the USA and some other English speaking countries this material is known as corn syrup. Despite the name the major ingredient is not dextrose but maltose. Throughout this work, to avoid confusion glucose is only used to refer to the syrup whereas chemical glucose is always referred to as dextrose.

Originally, glucose syrup was made by hydrolysing starch with acid. This process is controlled by measuring the proportion of the syrup that gives a Fehling's titration and assuming it to be dextrose. Thus, these syrups are specified in terms of 'dextrose equivalent', normally abbreviated to DE. Glucose syrup can be made from almost any source of carbohydrate but in practice it is only economic to produce glucose syrup from maize starch, potato starch or wheat starch – some wheat glucose is made as a by-product of the production of dried wheat gluten.

It is possible to take the process to completion to produce pure dextrose. This material obviously has a DE of 100. The commonest type of glucose syrup in sugar confectionery is 42 DE (or similar). This material is even referred to as confectioner's glucose. Other grades of glucose syrup are used in sugar confectionery, such as products of 68 DE or equivalent, which have the same water activity as invert sugar syrup and so can often be used as a direct replacement.

While glucose syrups were made by acid conversion, the DE gave a complete specification of the product. The ready availability of suitable enzymes has widened enormously the types of glucose syrups available.

Initially, syrups became available that were produced by an acid plus enzyme process, followed later by products that were produced completely enzymatically. The commercial advantage in this comes because a given weight of glucose syrup solids is cheaper than sucrose. The amount of sugar that can be replaced with glucose in a product is limited since 42 DE glucose is less sweet than sucrose and affects the water activity and other properties. The glucose industry started to use enzyme technology to produce high maltose glucose syrups. These products had the same DE as confectioner's glucose but because there was a higher proportion of maltose in the product the sweetness was higher, allowing more sucrose to be replaced by glucose. The technology of the glucose industry has now developed to the extent that virtually any starch hydrolysate can be produced if the demand is high enough.

The application of enzymes to glucose syrups was further extended to include the conversion of dextrose to fructose by glucose isomers. The resulting syrups were known as 'high fructose corn syrup' or isoglucose. The initial product was a syrup that was chemically equivalent to invert sugar syrup, and this product found a ready market in the soft drinks industry, particularly in the USA. (In Europe the authorities have not been keen on the idea of a product produced from starch, possibly of non-EU origin, replacing EU grown beet sugar.) As with dextrose, the conversion process can be continued to produce pure fructose.

Fructose

As already mentioned at the beginning of this chapter, fructose is normally encountered as a component of invert sugar. It has some properties that give rise to minor uses. Fructose is normally regarded as being twice as sweet as sucrose although high levels of fructose in a product tend to give a burning taste. One property of fructose which is sometimes useful is that, unlike other sugars, it is metabolised independently of insulin. For this reason fructose is sometimes used in products made specially for diabetics. It is claimed that small quantities of fructose smooth the taste of intense sweeteners when used in sucrose-free products. Although fructose can be made from glucose syrup by using glucose isomerase, in Europe the most common sources are found in chicory or Jerusalem artichokes.

Fructose is very soluble and is hence a very hygroscopic product – for this reason fructose is usually used as a syrup. Attempts to crystallise fructose by normal methods do not work, and for many years it was referred to as the uncrystallisable sugar. Fructose in a form which is described as crystalline is now available commercially and could well be produced by spray drying.

Dextrose

Pure dextrose is sometimes used as a confectionery ingredient and has roughly half the sweetness of sucrose. In Europe the use of dextrose is not particularly attractive commercially; however, in other parts of the world its use can be economically advantageous.

Lactose

Lactose, the major sugar found in milk, is a disaccharide reducing sugar, but unlike the other sugars it is not particularly soluble. Some individuals are unable to metabolise lactose and are therefore described as lactose intolerant. This is because they lack the enzyme lactase which is needed for lactose metabolism. Lactose intolerance is common in those parts of the world where humans do not consume any dairy products after weaning. In practice this means Asia, so it is possible that the majority of the world's population is lactose intolerant.

It is possible to produce lactose-removed skim milk. Another approach with lactose is to hydrolyse it to its constituent monosaccharides. As well as avoiding lactose intolerance this allows a syrup to be produced from cheese whey, and these syrups are offered as an ingredient for toffees and caramels.

Lactose is normally encountered as a component of any skim milk that is used in sugar confectionery but small quantities of crystalline lactose are also sometimes used in confectionery-making. If a product is made using too much lactose then a metallic taste appears, although the amount of lactose that can be consumed without this happening varies between individuals.

As one of the effects of the Common Agricultural Policy has been to increase the price of all milk products there has been some substitution of skim milk powder by products derived from whey. Impure grades of spray dried lactose derived from whey are offered as a confectionery ingredient.

DAIRY INGREDIENTS

Confectionery is not normally made directly from liquid milk as the amount of water that needs to be removed is too great. Milk solids are normally used as either milk powder or sweetened condensed milk.

Skim milk solids are an essential part of toffees and fudge – originally, full cream milk solids were used. Some products are still made using full cream milk solids but the majority now contain solids produced from skim milk. In some cases butter or butter oil is added to replace the fat that has been removed from the skim milk although it is possible for the fat content of the milk to be replaced with vegetable fat. It might appear curious that whole milk is effectively reconstituted from skim milk and butter but there are good reasons. Skim milk powder has the advantage over full cream milk powder in that it keeps better, and using skim milk and butter can, under certain conditions, be economically advantageous.

Sweetened Condensed Milk

The preferred source of milk solids in toffee manufacture remains sweetened condensed milk. This was one of the earliest ways of producing a stable product from milk, and nowadays both full cream and skim milk forms are used. The advantage of skimmed sweetened condensed milk is that the milk fat can be replaced with vegetable fat if so required. Toffees made from sweetened condensed milk are normally smoother than those made from milk powder: presumably the milk protein in sweetened condensed milk is in a less damaged form than in milk powder (see below). Sweetened condensed milk also has the advantage that provided the tin is not opened it keeps well without refrigeration. Sweetened condensed milk is a sticky syrup and needs some skill in handling; however, confectionery factories are expert in using sticky syrups.

Evaporated Milk (Unsweetened Condensed Milk)

Evaporated milk is a more modern product than sweetened condensed milk. It is not normally used to make confectionery as it tends to produce a taste in toffees that is not liked. This material has no technical or economic advantages over either sweetened condensed milk or milk powder.

Milk Powder

Milk powder (from both skim milk and whole milk) is the other form of milk solids used in confectionery. Skim milk was originally roller dried, but this process has now almost passed out of use and modern milk powders are made by spray drying. This does less damage to the proteins than the older roller process such that the bioavailability of the proteins in spray dried powder is higher than in the roller dried form. In confectionery-making this is not a particular advantage since confectionery products do not (or should not) form a major part of the diet.

The less severe heat treatment of modern milk powder production can lead to problems since enzymes present in the milk are not inactivated. The enzyme in milk products, particularly milk powder, that causes problems is lipase. It should be appreciated that this is not the native lipase of milk, but refers to bacterial lipases that have been produced during storage. Whereas the native lipase of milk is relatively easily deactivated, bacterial lipases are much more resistant to heat treatment. (Bulk cold storage of milk does seem to favour organisms that produce heat-resistant lipases.)

Lipase splits fatty acids from glycerol to produce free fatty acids, for example, butyric acid. If the original fat is butterfat then at low levels this produces a 'buttery' or 'creamy' flavour. As the free fatty acid content is increased, this strengthens the flavour to 'cheesy'. Normally in toffees free butyric acid is not a problem at any practical level, possibly because of losses during cooking. Other free fatty acids have different flavours. Lauric acid, which is found in nuts, tastes of soap. This is not too surprising as soap often contains sodium laurate. Lauric fat sources, such as hardened palm kernel oil, are often used as a substitute for butter; another potential source is nuts, which are sometimes combined with toffee. In any of these cases, lipolytic activity can shorten the shelf life of the product or render it totally unacceptable.

Butter

Butter is used principally as an ingredient of toffees and butterscotch. The manufacture of butter is one of the two oldest dairy processes, the other being cheese production. Traditionally, butter was made by allowing the cream to separate from the milk by standing the milk in shallow pans. The cream was then churned to produce a water in oil emulsion – typically, butter contains around 15% water. Butter is normally made as either sweet cream or lactic (also known as cultured), and with or without added salt. Lactic butter is made by adding a culture, usually a mixture of *Streptococcus cremoris*, *Str. diacetylactis* and *Betacoccus cremoris*. The culture produces lactic acid as well as various flavouring compounds, *e.g.* diacetyl which is commonly present at around 3 ppm. As well as any flavour effect, the lactic acid inhibits any undesirable microbiological activity in the aqueous phase of the butter. Sweet cream butter has no such culture added, but 1.5–3% salt is normally included: this inhibits microbiological problems by reducing the water activity of the aqueous phase. It is perfectly possible to make salted lactic butter or unsalted sweet cream butter if required. In the UK most butter is sweet cream whereas in continental Europe the lactic form is more common.

Another type of butter is whey butter, and this is produced from cream that has been skimmed off the whey after cheese-making. As the cream in whey butter has been subjected to the controlled lactic fermentation used in the cheese-making, whey butter has a characteristic and stronger flavour than other butters. This does not present any particular problems as any type of butter can be used to make toffee.

Traditionally, toffee makers preferred to buy rancid butter if available. As butter is stored, lipolysis causes the quantity of free fatty acids to rise. As mentioned in the previous section, one of these fatty acids, butyric acid, at low levels gives a pleasant, buttery flavour (butter flavours, in fact, tend to contain butyric acid). At higher levels the flavour becomes cheesy and at still higher levels takes on notes of Parmesan cheese. The response to butyric acid varies between individuals – some individuals regard butter as improving with storage. An approach that is used is to add a small quantity of lipolysed butter to the product. This has the same effect as using stale butter or adding a butter flavour. The lipolysed butter is a butter that has been deliberately treated with a lipolytic enzyme to release the fatty acids. One advantage of lipolysed butter is that it can be described as 'all natural', and depending upon the legislation is treated more kindly than a chemical flavour would be.

A common mistake about butter is the assumption that because it appears to be a solid then the fat must be crystalline. In fact, to crystallise completely all of the fat in butter it must be stored at $-40\,^{\circ}\mathrm{C}$. [This should be compared with the normal temperature of a deep freeze at -20 to $-18\,^{\circ}\mathrm{C}$ (*ca.* $0\,^{\circ}\mathrm{F}$).] As butter does not become completely liquid until 38–40 $^{\circ}\mathrm{C}$ this is an extremely wide crystallisation and melting range.

Unlike processed vegetable fat the composition of butter can only be altered by fractionation (typical fatty acid compositions of milk fats are given in Tables 3.5 and 3.6). It is possible to fractionate butter by a number of methods, either using solvents (such as acetone or alcohol) or by vacuum distillation or slow crystallisation. Solvent fractionation can be used to produce well defined fractions but has certain disadvantages:

Table 3.5 *Composition of unsaturated fatty acids in milk fat*

Monounsaturated acids		Polyunsaturated acids	
Carbon chain length	*Percentage*	*Carbon chain length*	*Percentage*
10:1	0.27	18:2	2.11
12:1	0.14	18:2 *cis, trans*-conjugated	0.63
14:1	0.76	18:2 *trans, trans*-conjugated	0.09
15:1	0.07	20:2	0.05
16:1	1.79	22:2	0.01
17:1	0.27	—	—
18:1	29.60	18:3	0.50
19:1	0.06	18:3 conjugated	0.01
20:1	0.22	20:3	0.11
21:1	0.02	22:3	—
22:1	0.03	—	—
23:1	0.03	20:4	0.14
24:1	0.01	22:4	0.05
—	—	20:5	0.04
—	—	22:5	0.06

Table 3.6 *Composition of straight and branched chain fatty acids in milk fat*

Straight chain acids		Branched chain acids	
Chain length	Percentage	Chain length	Percentage
4:0	2.79	—	—
5:0	0.01	—	—
6:0	2.34	—	—
7:0	0.02	—	—
8:0	1.06	—	—
9:0	0.03	—	—
10:0	3.04	—	—
11:0	0.03	—	—
12:0	2.87	—	—
13:0	0.06	13:0 br	0.04
14:0	8.94	14:0 br	0.10
15:0	0.79	15:0 brA, B	0.24, 0.38
16:0	23.80	16:0 br	0.17
17:0	0.70	17:0 brA, B	0.35, 0.25
18:0	13.20	18:0 br	Trace
19:0	0.27	20:0 br	Trace
20:0	0.28	—	—
21:0	0.04	—	—
22:0	0.11	—	—
23:0	0.03	—	—
24:0	0.07	—	—
25:0	0.01	—	—
26:0	0.07	—	—

even though the solvent residues are removed, volatile aroma compounds tend to get lost in the process. The original interest in fractionating butter came about to produce a butter product that would spread straight from the refrigerator, *i.e.* a butter to compete with soft margarine. Any fractionation process produces more than one fractionation product so a use had to be found for the hard fraction. Hard fats make it easier to make puff pastry. In some countries, pure butter puff pastry products, *e.g. mille feuilles*, are much appreciated. The hard fraction has turned out to be excellent for this purpose although, unfortunately, in spite of the early investigations into making a 'spreadable butter', demand for the soft fraction has not been sufficient.

Butter Oil (Anhydrous Milk Fat)

Butter oil is covered by an International Dairy Federation specification for anhydrous milk fat (see Table 3.7). Butter oil is milk fat with the water content reduced to 0.1% or less. It can be made by concentrating cream to 75% followed by treatment in a phase inverter before centrifugal separation, although it is more common to make butter oil

Table 3.7 *Specification[a] for butter oil from IDF standard 68a 1977 for anhydrous milk fat*

Milk fat	99.8% minimum
Moisture	0.1% maximum
Free fatty acids	0.3% maximum expressed as oleic acid
Copper	0.05 ppm maximum
Peroxide value	Not greater than 0.02[b]
Coliforms absent in 1 g	
Taste and odour	Clean, bland[c]
Neutralising substances	Absent

[a] This specification is for butter oil which is butter with the water removed. The free fatty acid limit is to detect lipolytic rancidity while peroxide value specification is to limit oxidative rancidity. The copper limit arises because copper catalyses the oxidation of fats. The absence of neutralising substances is to prevent a high titration for free fatty acids being covered up by the addition of alkali.
[b] Milli-equivalents of oxygen per kg of fat.
[c] Samples to be between 20 and 25 °C.

by melting the butter and removing the water with a centrifugal separator. At one time, butter oil was being made from butter that had been held in intervention stores. Owing to its low water content, butter oil has a very long shelf life and avoids the problems normally associated with butter storage. In some countries with no milk production, butter oil is combined with skim milk powder to produce milk products such as sweetened condensed milk, evaporated milk, ice cream and UHT milk.

Whey

Whey is the by-product of cheese-making. The traditional form of whey in confectionery is whey powder, which has been used as an ingredient in some toffees. Apart from this example, whey has not been much used in sugar confectionery. The reasons for this are hard to see, except that in toffees the flavours imparted by whey are not that pleasant. As the major ingredient is lactose this places another restriction on its use – lactose has its limited solubility compared with other sugars, when used to excess it imparts the unpleasant metallic taste, and it is not tolerated in the diets of certain consumers.

However, new technology has been applied to allow more whey to be used. Research has been carried out into converting the lactose to a mixture of dextrose and galactose. These two monosaccharides are both reducing sugars with the additional bonus that the mixture is much more soluble than lactose.

VEGETABLE FATS

Vegetable fats are mainly used in sugar confectionery as a substitute for milk fat; this is particularly so in the EU where the Common Agricul-

Table 3.8 *Approximate percentage fatty acid composition of fats and oils*

Source	C_4	C_6	C_8	C_{10}	C_{12}	C_{14}	C_{16}	$<C_{16}$ enoic	C_{16} enoic	C_{18}	$>C_{18}$	C_{18} enoic	$>C_{18}$ enoic	C_{18} dienoic	C_{18} trienoic
Butter fat	3–4	1–2	1–2	2–3	1–4	8–13	25–32	1–2	2–5	8–13	0.4–2.0	22–29	0.1–1.0	3	—
Coconut	—	—	—	4–10	44–51	13–18	7–10	—	—	1–4	—	5–8	0–1	1–3	—
Maize	—	—	—	—	—	0–2	8–10	—	1–2	1–4	—	30–50	0–2	34–56	—
Cottonseed	—	—	—	—	—	0–3	17–23	—	—	1–3	—	23–44	0–1	34–55	—
Olive oil	—	—	—	—	0–1	0–2	7–20	—	1–3	1–3	0–1	53–86	0–3	4–22	—
Palm oil	—	—	—	—	—	1–6	32–47	—	—	1–6	—	40–52	—	2–11	—
Palm kernel oil	—	—	2–4	3–7	45–52	14–19	6–9	—	0–1	1–3	1–2	10–18	—	1–2	—
Peanut	—	—	—	—	—	0.5	6–11	—	1–2	3–6	5–10	39–66	—	17–38	0–1
Soybean	—	—	—	—	—	0.3	7–11	—	0–1	2–5	1–3	22–34	—	50–60	2–10

tural Policy increased the price of milk fat (see Table 3.8 for a comparison of the compositions of butter and vegetable fats). Because vegetable fats can be blended, hydrogenated and interesterified it is possible to produce a vegetable fat with almost any desired range of properties. In toffees the fats that are used as an alternative to milk fat are not an attempt to match the composition of milk fat but are designed to provide the best blend of properties for the product.

The situation was not always as it is now. The original vegetable fat used in toffees was hardened palm kernel oil (HPKO). This material does have the advantage of being cheap but, unfortunately, it is a lauric fat and tends to reduce the shelf life because of soapy rancidity.

GUMS AND GELLING AGENTS OR HYDROCOLLOIDS

Legally speaking, another title which some of these ingredients fall under is that of thickeners and stabilisers. Some are only minor components of confectionery and can properly be regarded as additives; others are used in quantities which make them ingredients.

Gelling agents under appropriate conditions self-associate to produce a three-dimensional structure. Some gelling, as with gelatine, is thermo-reversible; other gelling, such as with high methoxyl pectin, is irreversible. Apart from the effects on the texture of the product, an irreversible gelling agent is more of a problem in the factory since it cannot readily be recycled.

The individual substances, their origins and properties are given below and are tabulated for comparison in Table 3.9.

Agar Agar, E406

The name is not a printer's error although the substance is often referred to as agar. This gelling agent is a seaweed polysaccharide and it is extracted from red seaweeds from Japan, New Zealand, Denmark, Australia, South Africa and Spain although there are other possible sources. Initially, the agar is extracted from the seaweed with hot water and the resulting solution concentrated: typical methods are freezing and thawing or heating under vacuum. The vacuum reduces the boiling point which both saves energy and reduces damage to the agar. The commercial form of agar can be either strips, flakes or powder, the colour of which may be improved by the use of bleach. The finished product can have a characteristic flavour and odour. Typically the molecular weight is over 20 000.

The strength of gel given by agar varies depending upon its origin and this makes it necessary to test batches of agar for gel strength before use. As the strength of the gel is also dependent upon the pH and total solids,

Table 3.9 *Properties, chemistry and sources for gums and gelling agents*

Agent	Properties	Chemistry	Source
Gelatine	Thermoreversible gelling agent	Protein	Bovine or porcine hides or bones
Starch	Irreversible gelling agent	Carbohydrate	Maize, wheat or potatoes
High amylopectin starch	Non-gelling starch	Carbohydrate	Waxy maize
Gum acacia	Gum	Polysaccharide	Trees of the species *Acacia senegal*
Agar agar	Thermoreversible gelling agent	Polysaccharide	Red seaweeds
Alginate	Irreversible gelling agent	Polysaccharide	Brown seaweeds
Carrageenan	Thermoreversible gelling agent	Sulfated polysaccharide	Red seaweeds
Gellan gum	Thermoreversible or irreversible gelling agent	Polysaccharide	*Pseudomonas elodea*
Guar gum	Thickener exhibits synergy with some gelling agents	Galactomannan	Seeds of *Cyamopsis tetragonolobus*
Pectin, high methoxyl	Irreversible gelling agent	Polygalacturonic acid	Citrus peel or cider apple pomace
Pectin, low methoxyl	Thermoreversible gelling agent	De-methoxylated pectin	Citrus peel or cider apple pomace
Gum tragacanth	Gum or mucilage	Polysaccharide	*Astragalus* shrub
Locust bean or carob gum	Thickener exhibits synergy with some gelling agents	Galactomannan	Endosperm of locust beans from *Ceratonia siliqua*
Xanthan gum	Thickener exhibits synergy with locust bean gum	Polysaccharide	Aerobic fermentation of *Xanthomonas campestris*
Egg albumen	Whipping agent and irreversible gelling agent	Protein	Egg white
Enzyme modified soy protein	Whipping agent	Protein	Soy beans

the strength test needs to be performed under the conditions of those used in the manufacturing process. Agar is not considered to carry fruit flavours well, although if used in fruit jellies a typical pH is between 4.5 and 5.5.

In use, agar solution is normally prepared by mixing agar with ten times its own weight of sugar and then dissolving it in 30–50 times its own weight of water, the sugar being added to prevent the agar forming lumps upon addition to the water. This technique is often used when adding a slow-dissolving solid to water as the sugar dissolves quickly, helping to disperse the agar. Alternatively, agar can be added to boiling water. Temperatures above 90 °C (194 °F) are needed to dissolve agar, although agar is resistant to heat unless the conditions are acid. Naturally, for reasons of safety it is normal to reduce the temperature to 60 °C before adding any acid.

The gel strength is not directly related to the proportion of gelling agent used, the maximum gel strength being obtained at pH 8–9 with the solids between 76 and 78% – if the solids are above 80% the gel strength is reduced. In a typical confectionery fruit gel, around 0.5–1.5% agar is used and the typical texture of an agar jelly is described as 'short'. The jelly texture may be altered by adding jams or fruit pulps, where textural changes are likely to be brought about by the pectins that occur naturally within the fruit. The gel strength of agar jellies is increased by adding locust bean gum but is reduced by adding alginates or starch.

Compared with other gelling agents agar needs more water to get it into solution. Agar gels display hysteresis, in melting at 85–90 °C (185–194 °F) but setting at 30–40 °C (86–104 °F). This is a useful property when the product is being deposited hot into moulds as there is no possibility of premature setting or pre-gelling. It takes longer for an agar gel to set in starch moulds than it does for an equivalent product made from pectin. Agar is more difficult to handle in automated plants than some equivalent products. One advantage of agar is that it is not metabolised in the body, which does lead to a slight reduction in energy content of the overall product.

Alginates, E401

These materials are polysaccharides and the name derives from their original source, brown algae. Current commercial sources are brown seaweeds such as *Laminaria digitata*, *Laminaria hyperborea*, *Ascophyllium nodosum* and *Fucus serratus*. Different properties are obtained in alginates from different seaweeds found in different regions, where typical sources are found along rocky coastlines in the US, the UK, France and Norway.

The seaweed must first be treated with acid to convert the alginates to alginic acid which is insoluble. The next stage removes the soluble impurities such as mannitol or mineral salts leaving the alginic acid in the purified seaweed. Treating the seaweed with alkali converts the alginic acid to alginates which are soluble in alkali. The insolubles, *e.g.* cellulosic and proteinaceous materials, are removed at this stage by filtration, flotation and settling. The alginate is then precipitated with acid as alginic acid and the precipitate is washed and dried. The desired alginate is then produced by adding alkali. The product is subsequently dry milled and sieved to the required particle size.

The monomers of alginate are mannuronic (M; **1**) and guluronic (G; **2**) acids, the polymers being composed of three types of sequence: homo-

geneous –M–M– segments, homogeneous –G–G– segments and hetero-
geneous –M–G–M–G– segments.

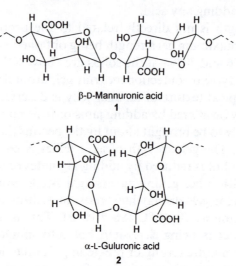

β-D-Mannuronic acid
1

α-L-Guluronic acid
2

Alginates are soluble in alkaline solutions but insoluble in aqueous
acids and organic solvents. In cold water, alkaline alginates only form
gels in the presence of calcium, where, for the guluronic acid segments
only, the gel forms an 'egg box' structure (Figure 3.1). Calcium alginate
has the formula $[(C_6H_7O_6)_2Ca]_n$ and has a molecular weight in the range
32 000–250 000. Gels formed by calcium alginate are not thermoreversible but the gelling ability is proportional to the amount of homogeneous G segments present. In the absence of calcium ions, alginates act
only as thickeners by increasing the viscosity of the solution.

Alginates are little-used in confectionery, but one suggested application is as a gloss and non-stick coating on liquorice products – the
traditional gloss, mineral oil, has been banned from food use.

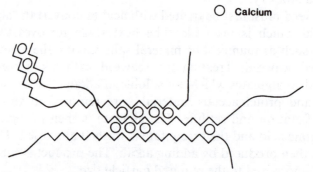

○ Calcium

Figure 3.1   *Egg box-type bonding*
(Reprinted with permission from E.B. Jackson, *Sugar Confectionery Manufacture*, Figure 3-2, p. 45, © 1995, 1999 Aspen Publishers Inc.)

Carrageenan

This material is another seaweed polysaccharide. The name is believed to be of Irish origin. The ability of certain seaweeds to gel large quantities of milk was noted in a number of places including the West of Ireland and Brittany – the Breton milk gel was called blancmange.

The commercial sources of carrageenan include the following red seaweeds: *Euchema cottonii*, *Euchema spinosum*, *Chondrus crispus*, *Gigartina acicularis*, *Gigartina stellata*, *Gigartina pistallata*, *Gigartina skottsbergii*, *Gigartina chamissoi* and *Iradaea*. These seaweeds grow in Argentina, France, Morocco, Peru, Chile, The Philippines and Indonesia.

Carrageenan is extracted by first treating washed seaweed with hot water. The seaweed is then crushed in the presence of alkali to extract the maximum amount of carrageeenan. A diatomaceous earth is then used as a filter aid in filtering the hot extract under pressure. The product produced at the end of this stage is a clear syrup. The carrageenan is then precipitated from this syrup with alcohol. The coagulated carrageenan forms into fibres which are then pressed and washed with strong alcohol to dehydrate them.

Ethyl alcohol forms an azeotrope with water. This product is the rectified spirit found in laboratories. It is not possible to produce a higher concentration of alcohol than this without distilling at reduced pressure or using other additives. The strong alcohol used to dehydrate the carrageenan is like the absolute alcohol found in laboratories; it has less water than the azeotrope. When the carrageenan is treated with the strong alcohol the alcohol extracts the residual water to bring the water content of the alcohol towards the azeotrope. The alcohol can then be recycled by vacuuming distilling out the water. The alcohol used must be food grade. In the UK, duty would be charged on this material.

Carrageenan is chemically a sulfated polysaccharide consisting of galactose units. A common backbone exists in all the different fractions. The main chain is composed of D-galactose residues linked alternately $\alpha(1\rightarrow3)$ and $\beta(1\rightarrow4)$. The fractions are distinguished by the different number and position of the sulfate groups. A 3,6-anhydro-bridge can exist on the galactose linked through the 1- and 4-positions. The gelling carrageenans κ and ι contain β-D-galactose 4-sulfate linked through the 1- and 4-positions. κ- and ι-Carrageenan differ in that ι-carrageenan contains an additional sulfate group on the 3,6-anhydrogalactose. These carrageenans are always found contaminated with the other. μ- and ν-Carrageenan are the biological precursors of κ- and ι-carrageenan. The seaweeds have an enzyme that catalyses the transformation by eliminating the 6-sulfate group. Conveniently, the alkaline extraction used on carrageenan also expedites this transformation thereby improving the quality of the product as a gelling agent.

Different species of seaweed yield different carrageenan fractions:

κ-carrageenan, found in *Euchema cottonii*;
ι-carrageenan, found in *Euchema spinosum*;
λ-carrageenan, found in *Gigartina acicularis*.

κ-, ι- and λ-Carrageenan are found in *Chondrus crispus*, *Gigartina stellata* and *Iradaea*.

κ-Carrageenan is a gelling agent that is soluble at 60–70 °C. It reacts producing a thermoreversible gel with milk protein. The gels tend to syneresis and tend to be breakable. A synergy exists between carrageenan and carob gum, *i.e.* the two together have more effect than both used singly.

ι-Carrageenan dissolves at around 55 °C and is a gelling agent. It reacts with milk protein to give elastic gels that do not undergo syneresis. The gels formed from ι-carrageenan are thermoreversible. In contrast, λ-carrageenan is cold soluble and is used as a thickener. The gels that it gives with milk protein are very weak.

When carrageenan forms gels the chains associate through double helices. Any μ- or ν-carrageenan present inhibits this process.

Gelatine

Gelatine is one of the most versatile sugar confectionery ingredients and is produced by hydrolysing collagen (tropocollagen), a connective protein found in the bones and hides of animals. Commercial sources of gelatine are normally cattle or pigs. Collagen consists of three polypeptide chains arranged in a triple helix and can be hydrolysed under either acidic or alkaline conditions. (Gelatine prepared by acid hydrolysis is normally referred to as type A whereas gelatine produced by alkaline hydrolysis is referred to as type B.) In contrast, the structure of gelatine consists of a number of free or inter-associated chains ranging in molecular weight from around ten thousand to several hundred thousand. On extraction, monomers (α-chains MW 100 000), dimers (β-chains), trimers (λ-chains) and some lower order peptides are released.

In use, gelatine is pre-soaked whereupon it absorbs five to ten times its own weight of water. The swollen gelatine will then dissolve at 50–60 °C. As gelatine can be hydrolysed by heating it to above 80 °C, trying to dissolve gelatine by boiling it (or boiling a gelatine solution) is inadvisable as it simply hydrolyses the gelatine further. As gelatine is not stable to acid, any addition of such must be as late as possible in the process.

A very important property of proteins is the isoelectric point. This is defined as the point at which the total negative and positive charges on the molecule are balanced. This point is where it is easiest to precipitate the protein. By analogy with the pH scale the isoelectric point is written pI, where

type A, acid-processed, gelatine, pI = 6.3–9.5;
type B, alkali-processed, gelatine, pI = 4.5–5.2.

Different types of gelatine cannot be mixed as they have different isoelectric points and these different types are referred to by their origin and the agent used for hydrolysis, *e.g.* acid pigskin gelatine. Limed ossein gelatine is normally produced from cattle bones.

In producing gelatine, skins are processed directly, whereas the bones are washed to remove the meat and fat residues which produces dry, degreased bone. The bones are then treated with hydrochloric acid which removes the phosphates that are present. These are then precipitated with lime to produce di-calcium phosphate for use in cattle food. The remainder of the bones is the bone or ossein collagen. The collagen, whether bone or hide, is then hydrolysed with acid or alkali (usually lime) as required. In the alkaline hydrolysis process the collagen is steeped in a bath of lime for a number of weeks at ambient temperature. In comparison, acid hydrolysis takes only one day at ambient temperature. Next, the acid or alkali is washed away and any remainder is neutralised before the material is then cooked with hot water in order to liberate the gelatine. The resulting solution contains about 6–7% gelatine and is referred to as an 'extraction'. Repeated extractions are made until the raw material is exhausted – the highest quality gelatine is in the earliest extractions. (If the starting material was pig skin the fat can be recovered at this stage.) The extractions are then filtered followed by vacuum evaporation to 30–40% gelatine. Next, the solution is sterilised at 140 °C followed by crash cooling to obtain a jelly – the crash cooling minimises degradation of the gelatine. The concentrated gelatine jelly is then extruded using extruders similar to those used to make pasta. The extruded material is then air dried to the final moisture content before the dried material is ground as needed followed by blending to the required specification. A typical product is 14% moisture, 84% protein and 2% ash.

As gelatine is hygroscopic it needs to be stored so that it cannot pick up water. If the moisture content is allowed to rise to 16%, mould growth can commence. Care needs to be taken when working with gelatine solutions as they form an excellent medium for bacterial growth. Gelatine is a product where the microbiological quality is very important. The performance of the gelatine is impaired if it suffers bacterial proteolysis; more importantly, as an animal product there is always a risk of contamination with pathogenic bacteria.

As gelatine is an animal product it is unacceptable to vegetarians. Some religious groups also have problems with it although Kosher gelatine is available. Gelatine made from fish has recently become commercially available.

The gelling quality of gelatine is obviously an important property to users, and is normally measured using a number of non-SI empirical methods. Measures in use are grams Bloom, Boucher units, FIRA degrees and jelly strength. Of these the most used is Bloom strength. It is possible to use either a smaller quantity of a high Bloom gelatine or a greater amount of a low Bloom gelatine to produce the same result. Table 3.10 shows the percentage of gelatine of other Bloom grades to give a similar strength to 100 Bloom gelatine and Table 3.11 gives the

Table 3.10 *Relationship between a solution of 100 Bloom strength gelatine and equivalent jelly strength of other Bloom grades. Values are concentration (%) of gelatine required to give a similar jelly strength.*

Bloom strength of gelatine (%)

60	80	100	140	160	200	225	260
7.7	6.7	6.0	5.1	4.8	4.3	4.0	3.7
10.3	8.9	8.0	6.8	6.3	5.7	5.3	5.0
12.9	11.2	10.0	8.4	7.9	7.1	6.7	6.2
15.5	13.4	12.0	10.1	9.5	8.5	8.0	7.4
18.1	15.7	14.0	11.8	11.1	9.9	9.3	8.7

(Taken from *Sugar Confectionery and Chocolate Manufacture*, R. Lees and B. Jackson (1973) Leonard Hill, Glasgow)

Table 3.11 *Relationship between Bloom strength of various gelatines held in solution having equivalent jelly strength*

Bloom strength of gelatine (%)	Gelatine needed to produce equivalent jelly strength (%)				
	60 Bloom	100 Bloom	160 Bloom	200 Bloom	260 Bloom
60	10.0	12.9	14.9	18.3	20.7
80	8.6	11.2	14.2	15.8	18.0
100	7.8	10.0	12.6	14.1	16.1
120	7.1	9.1	11.5	12.9	14.7
140	6.6	8.4	10.7	12.0	13.6
160	6.1	7.9	10.0	11.2	12.8
180	—	7.5	9.4	10.5	12.0
200	—	7.1	8.9	10.0	11.4
220	—	6.8	8.5	9.5	10.9
240	—	6.5	8.1	9.1	10.4
260	—	6.2	7.8	8.7	10.0

(Taken from *Sugar Confectionery and Chocolate Manufacture*, R. Lees and B. Jackson (1973) Leonard Hill, Glasgow)

relationship between Bloom strengths and concentration for equivalent jelly strength. Commonly available Bloom strengths are 60–260 although higher strengths are available.

Gelatine has the property that on standing it can set to acquire the structure of the original collagen. Gelatine gels form provided that the concentration is high enough and the temperature is low enough. Thus, for any concentration of gelatine gel there will be a setting temperature. The thermoreversible nature of gelatine gels is useful in several ways. The product can give a melt-in-the-mouth sensation, waste material can be recycled and it is possible to deposit the product hot and leave it to set on cooling.

As well as its use as a gelling agent gelatine can be used as a foaming agent. Proteins tend to stabilise foams. When a mixture containing gelatine is whipped it is possible to arrange that the mixture cools thus setting the foam.

Gelatine produces some textures which cannot be produced otherwise. For example, the photographic industry has been unable to find a substitute for gelatine in photographic film and paper. Gelatine also has some uses in pharmaceutical products, *e.g.* gelatine capsules. Table 3.12 gives some uses of gelatine in confectionery with the type and percentage of gelatine used. Gelatine does have a few minor uses in confectionery such as sealing almonds in sugared almonds and as a granulation binder for pressed sugar tablets.

It is possible to use gelatine in combination with other hydrocolloids such as pectin, agar, starch or gum acacia – gelatine and gum acacia have been used in Rowntree's fruit pastilles for over 100 years. However, using a mixture of hydrocolloids can lead to difficulties. In the case of gum acacia and gelatine, if the conditions are wrong then co-acervates will form. Under various conditions, macromolecular solutions may separate into two liquid layers, one poor in colloid the other rich in it. In colloid science this partial miscibility has been called *coacervation* and the colloid rich phase *coacervate* (see H.R. Kruyt, *Colloid Science*, Vol. II, Elsevier, New York, 1949). Using a mixture of hydrocolloids allows a range of textures to be produced. As examples, gelatine and gum give a hard

Table 3.12 *The uses and properties of gelatine in confectionery*

Product	Bloom	Property used	% gelatine
Jellies	175–250	Gelling agent	6–9
Wine gums	100–150	Gelling agent	4–8
Marshmallow	200–250	Whipping agent	2–5
Fruit chews	100–150	Whipping agent	0.5–2.5
Extruded aerated products	101–125	Whipping agent	3–7

(Reprinted with permission from E.B. Jackson, *Suger Confectionery Manufacture*, Table 3-3, p. 49, © 1995, 1999 Aspen Publishers Inc.)

compact texture while a gelatine agar and pectin mixture will give a short brittle texture. Gelatine and starch give a texture between these extremes.

Gellan Gum, E418

Gellan gum has only relatively recently been introduced as a gelling agent and at the time of writing it is not universally legal in foods. The substance is the extra cellular polysaccharide produced in the aerobic fermentation of *Pseudomonas elodea*. The organism is fed a carbo-hydrate, *e.g.* glucose, with a nitrogen source and inorganic salts. The production system works under very carefully controlled conditions of aeration, pH and temperature, and the broth produced is treated with hot alkali before the gum is precipitated by treating with hot propan-2-ol. The monomers in gellan gum are rhamnose, glucose and glucuronic acid in the ratio $1:2:1$.

Gellan gum has been promoted as a suitable gelling agent for making fruit flavour jellies. The flavour release has been claimed to be excellent. It is particularly suited to this application as it is very stable even in acid conditions.

The gels are made by adding the gellan gum to water while shearing followed by heating to $75\,^{\circ}$C then adding ions and allowing to set by cooling. The level of gellan gum needed to form a gel can be as low as 0.05%. As gellan gum sets in the presence of ions, suitable salts must be present. Suitable salts permitted in foods are those of potassium, calcium, sodium or magnesium, and the divalent cations such as calcium and magnesium will gel the gellan gum at $\frac{1}{25}$th of the amount of the monovalent cations sodium or potassium.

It is possible to make clear confectionery gums from gellan gum, and it is also possible to make gellan gums that do or do not melt on heating, as required. The texture produced is dependent upon the concentration of the gum, the total soluble solids, the ion concentration and the pH, although the texture of the gels produced does tend to be brittle. One way of modifying this is to add other gelling agents and it is claimed that gellan gum is compatible with gelatine, xanthan gum, locust bean gum and starch.

Although it is not universally legal in foods, gellan gum has been approved for food use in a number of countries – it will probably be approved everywhere within the next few years – but in spite of the studies made of gellan gum it is not yet used in confectionery. The problem is undoubtedly that it will not make the same product as another gelling agent does for confectionery already on the market. The prospect for gellan gum therefore depends on it being used in a new product; however, successful new products are few and far between in confectionery. If the alkali step is removed a high acetyl gellan gum is produced – this type of gellan gum gives a more elastic and thermo-reversible gel similar to gelatine. This product might be more successful but it needs separate food approvals.

Gum Acacia, also known as Gum Arabic, E414

Gum acacia is the exudate of *Acacia senegal* trees. The trees grow on the edges of deserts where they prevent the advance of the desert boundary by binding the sand together and retaining moisture. Because the trees are part of the genus Leguminosae they fix some nitrogen into the soil. Properly managed, the trees provide a cash crop and a source of firewood in addition to maintaining the fertility of the soil. Most gum is cultivated but a little comes from wild trees. Traditionally, the best gum comes from the area around Kordofan in the Sudan. In fact, most gum acacia comes from the Sudan, although other countries that produce acacia gums are those such as Chad, Senegal and Niger.

Gum acacia is a unique polysaccharide with some peptides as part of its structure. It has a range of different uses in confectionery (see Table 3.13). Originally, it was the gum in gum sweets although some gum sweets do now contain modified starch as a substitute. The replacement of gum is not because the substitute performs better but because there have been supply problems with gum acacia. In confectionery, the light-coloured grades are used to make products that need a light colour whereas darker gum is used in products that have dark colours and flavours, *e.g.* liquorice.

Traditionally, the best gum was produced by hand-picking clear tears of gum. This grade is still available but the price reflects the cost of the hand-picking. As the gum is produced by removing the tears from the gum trees some gum is contaminated with pieces of bark. Because of this the gum can pick up colour and astringent tastes from the bark. Also, raw gum is often contaminated with desert sand.

Table 3.13 *Uses of gum acacia*

Product	Use	Comment
Gum sweets	Bulk ingredient	A major proportion of these sweets will be gum
Pastilles	Bulk ingredient	The gum is used with gelatine in a lower proportion than in gums used in very hard pastilles, unlike the French and British products
Panning	Sealing between layers	Only a small proportion of the finished product is gum. Gum is used to form a barrier between layers, *e.g.* sealing almonds prior to sugar panning
Tabletting	Granule binder	A minor use of gum
Sugar-free boiled sweets	Stabiliser	A new use but as a minor ingredient. The high molecular weight of the gum prevents cold flow

There is a need to have a testing regime to ensure that gum acacia offered is gum acacia and not a product from some other species that is unsuitable. Acacia Seyal gum is sometimes encountered which is less soluble than gum acacia, and hence unsuitable for making sweets requiring a high proportion of gum acacia as it will not dissolve sufficiently. Instances have occurred where gum combretum, a product that is not an acacia gum, has been found in commercial supplies purporting to be gum acacia.

As raw gum acacia tends to arrive with natural contamination by bark and sand it is normally purified by filtering and centrifuging in order to remove any insoluble material. The gum is then dissolved in water with gentle heating to produce a solution of around 30–50% gum. Gum acacia is much more soluble than other gums and it is possible to make a 50% solution in cold water if needed. The solution viscosity falls with increasing temperature as well as being pH dependent. Maximum viscosity occurs at pH 6 but falls above pH 9 and below pH 4.

Some gum users now take gum in a pre-prepared form. An example is spray dried gum acacia which has been used in pharmaceutical products for some time – this offers the pharmaceutical manufacturer a clean and ready-to-use product. Holding the same advantages, instant forms of gum acacia also have been offered to confectioners for some time although, obviously, the instant gum is more expensive. A confectionery manufacturer that uses gum as a minor ingredient may well find that the capital and labour cost of purifying the raw material is not cost effective. However, a company that uses gum acacia as a major ingredient might come to a different conclusion.

Instantised gums pose different problems to the analytical chemist. One approach that can be used is to have an optical rotation specification for the product although this approach is not entirely proof against a material that contains a blend of gums of different optical rotations.

Guar Gum

This vegetable gum comes from *Cyamopsis tetragonolopus*, extracted from the endosperm of the seeds. The countries of origin are India and Pakistan.

Guar flour is made by milling the endosperms, having first de-hulled them. The resulting product is obviously impure and gives a cloudy aqueous solution.

The pure product is produced by dissolving the gum from the seeds in hot water. Diatomaceous earth filtration is then used to purify the solution. As the gum is less soluble in alcoholic than aqueous solutions it is precipitated by adding propan-2-ol. The pressed filter cake is then washed in pure alcohol to dehydrate it. The alcohol is then recovered by pressing again before the product is milled to the required final size.

Chemically, guar gum is a galactomannan, *i.e.* it is composed of β-D-mannose and α-D-galactose units. The molecule has a main chain composed of $(1\rightarrow4)$-linked β-D-mannose residues with side chains of $(1\rightarrow6)$-linked α-D-galactose. Guar gum is chemically very similar to locust bean gum and was originally developed to make up for locust bean gum shortages. The differences between the two gums are in the number of galactose molecules attached to the mannose chain. Guar gum has a ratio of D-galactose to D-mannose of 1:2 whereas the same ratio in locust bean gum is 1:4. While locust bean gum does not itself gel it does form gels with carrageenan. Locust bean gum is a minor ingredient in confectionery.

Pectin

Pectin is used in confectionery in two forms, high methoxyl pectin and low methoxyl pectin. High methoxyl pectin is the form normally found in fruit whereas the low methoxyl form is a chemically modified pectin. Pectins are acidic polysaccharides that occur in the cell walls of fruit. The commercial source of pectin is either citrus peel or apple pomace, the citrus peel being the residue from the production of citrus juices and apple pomace the residue of cider production. Thus pectin is a by-product of either cider or fruit juice production.

Pectins are extracted from the raw material by using hot hydrochloric acid. The protopectin present in the fruit is hydrolysed in the acid treatment and then pressing and filtration remove the insoluble material present. The next stage in the process is to precipitate the pectin and the way in which this is done depends upon the type of pectin being manufactured. A rapid set, high methoxyl pectin is precipitated as soon as possible, whereas a lower degree of methoxylation is obtained by holding the extract for several days, which removes some of the methoxyl groups. (If amidated low meythoxyl pectins are being made then the pectin is treated with ammonia at this stage.) The pectin is then precipitated with alcohol and the resulting precipitate is washed with alcohol of successively higher strength finishing with alcohol of pure grade. The pure alcohol removes the residual water as it has a higher affinity for water than the pectin. This gives a fibrous pectin which is then dried, ground and sieved. An alternative method of precipitating pectin is to treat it with aluminium ions in order to produce an insoluble aluminium pectin salt. The aluminium is subsequently removed by treating with acidified alcohol.

Chemically, pectins can be regarded as a polygalacturonic acid – the pectin molecule is a polymer with galacturonic acid (**3**) monomers linked through (1→4) bonds. Typical molecular weights are between 2000 and 100 000. Some of the galacturonic acid groups will be methoxylated (**4**) and the ratio of methoxylated to unmethoxylated galacturonic acids is referred to as the degree of methylation, normally abbreviated to DM. This is an important parameter as it controls how the pectin behaves as well as how it is treated in food legislation. The DM is defined as the average number of methoxyl units per 100 galacturonic units. Pectins with a DM above 50 are referred to as high methoxyl whereas those with a DM below 50 are classed as the low methoxyl forms. The low methoxyl pectins are sometimes further modified by converting some of the acid groups to amides by treating the pectin with ammonia. As with methoxylation, the degree of amidation is the average number of amide groups per 100 galacturonic units. Most countries restrict the degree of amidation to a maximum of 25%. High methoxyl pectins are naturally present in fruit and escape restrictions on use for that reason. Low methoxyl pectins are, however, treated as additives and have restrictive acceptable daily intakes (ADIs).

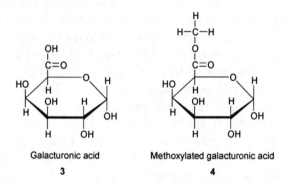

Galacturonic acid   Methoxylated galacturonic acid

3   **4**

    As the setting conditions for high methoxyl pectins require high soluble solids and a low pH, high methoxyl pectins are used for making low pH products such as fruit jellies. The high methoxyl pectins are further subdivided by their speed of gelation. The speed of gelation is controlled by the DM. A DM around 72% gives a rapid set while a DM around 68% gives a medium set. A slow set is achieved with a DM below 64%. In confectionery products the soluble solids are so high that only slow set pectins are used since a rapid set pectin would pre-gel under these conditions.

    Low methoxyl pectins have radically different properties to those of the high methoxyl forms – the small chemical modification has totally altered the way in which the pectins behave. The gelling conditions for low methoxyl pectin require a pH of 2.8–6.5 (outside of which range the product would be inedible) and between 10 and 80% soluble solids in the presence of calcium.

The gel properties, as well as the gelling conditions, are radically different for the two types of pectin. High methoxyl pectins produce a gel that does not remelt, whereas some low methoxyl pectin gels are thermoreversible as well as having a softer and less elastic texture.

When making fruit-flavoured jellies the normal type of pectin to use is high methoxyl, and as high methoxyl pectins require an acid pH to set these products are normally acidified as part of the fruit flavour. When making products with a neutral flavour, for example vanilla- and mint-flavoured jellies and rose water (for use in Turkish delight), a neutral pH is also commonly found (pH *ca.* 5), making the use of high methoxyl pectins unsuitable here. The only option, therefore, is to use a low methoxyl pectin.

The ability to trigger the gelling mechanism by adding calcium ions has led to several innovative ideas: one is to use the gel to make crustless liqueurs; another is to make liquid-centred fruit-flavoured products. However, because hard water normally contains calcium ions, care must be taken in selecting low methoxyl pectins when using hard water supplies or when moving recipes between factories. The gel produced by low methoxyl pectins has the egg-box structure also found in alginates (Figure 3.1).

Pectin does have some compatibility with other gelling agents. In particular it is used in conjunction with gelatine where higher levels of replacement by pectin cause the texture to become softer and less chewy. Pectin suppliers claim that up to 25% of the gelatine can be replaced without significantly altering the texture. The benefit of the replacement is, of course, purely financial. Pectin is also compatible with starch but gives a pasty texture which is not very popular although it does seem to be acceptable in Turkish delight.

The problem with mixing hydrocolloids is that if the pH is wrong then the entire system becomes unstable. The stability of the hydrocolloid in solution depends on the electrostatic charge on the molecules, and anything that neutralises that charge is likely to bring the material out of solution.

The method for using high methoxyl pectins is as follows:

(1) The pectin is dissolved in water with one third of the acid. If a high shear mixer is available then the pectin can be added directly to the water although if this is tried without a high shear mixer then the pectin will form blobs of jelly. This can be avoided by dry mixing the pectin with five times its own weight of sucrose – when the mixture is added to water the sugar prevents the pectin gelling with itself.
(2) The solution is boiled to dissolve the pectin completely and the remaining sugar and glucose syrup are added.

(3) The solution is then further heated to boiling point to dissolve all of the sugar and is concentrated by boiling to the desired final solids content.
(4) The colours and flavours and remaining acid are then added. Once this acid has been added the product can set.
(5) The product is then deposited into moulds.
(6) The moulds are allowed to stand to allow the product to cool and set.

As pectin gels without being further concentrated, pectin-based products can be shaped using rubber moulds. In the terminology of the industry this is 'starchless moulding'. Starch moulding used depressions in starch as moulds – some pectin jellies are still made using this method. Starch moulding has the advantage of greater flexibility and lower capital cost. This is because in a starchless system, changing the product shape requires a whole new set of moulds, whereas in a starch-based system all that is required is a new mould board to stamp a different shape in the starch.

In working with high methoxyl pectin the pH must be controlled because below pH 4.5 slow set pectins degrade causing a loss in gel strength. At pH below 3.2 there is a danger that the pectin will pre-gel when the solution reaches the final strength. It is normal to use a buffer, such as citric acid and potassium citrate, to prevent these problems. Pectins premixed with suitable buffer salts are commercially available. This type of product avoids having staff skilled enough to make up buffers accurately and is attractive to small manufacturers.

One problem with high methoxyl pectin is that of using rework. As pectin gels do not remelt this is much harder to work with than it is with gelatine. This point is important because not only is some of an expensive ingredient lost, but simply because it also generates the additional cost for waste disposal. In practice, high methoxyl pectin rework can be used if it is first comminuted, but the proportion used must not exceed 5%.

Pectin has been suggested as an ingredient for aerated products where a compatible whipping agent must also be used in conjunction with a high methoxyl pectin. Typically the product will contain 0.5–2.5% of high methoxyl pectin and some gelatine.

Starch

Starch is the major energy storage polysaccharide of cereal crops. It is a natural polymer of dextrose. Starch has two naturally-occurring forms: one is amylose, a polymer with long linear chains; the other is amylopectin which is a branched chain polymer. The length of the amylose chain varies between different plants but common values are between 200

Table 3.14 *Starch sources, types and uses*

Source	Type of starch	Typical starch product	Confectionery uses
Maize	Standard	Acid-thinned boiling starch	Gums, jellies, pastilles
Potato	Standard	Acid-thinned boiling starch	Gums, jellies, pastilles
Wheat	Standard	Acid-thinned boiling starch	Gums, jellies, pastilles
Waxy maize	Speciality	Oxidised waxy maize	Sealing panned centres
Tapioca	Starch ether	Hydroxypropylated and thinned starch	Gums
Tapioca	Speciality native starch	Native starch	Sealant for panned products

and 500 glucose units. Although the chain lengths in amylopectin are limited to 20–30 glucose units it tends to be a more massive molecule than amylose. Different plants have different ratios of amylose to amylopectin; indeed this is responsible for most of the variations in properties between starches of different plants. Virtually any starch-containing crop can be used as a source of starch (see Table 3.14). The sources that are used commercially are maize (US corn), wheat, potato, rice tapioca, or sago. The common types of starch are extracted from maize, wheat or potatoes. The other crops are used when starches with special properties are required. The choice of raw material ultimately depends upon economics and availability. Wheat starch can be produced as a by-product of the production of dried wheat gluten.

The methods used to separate the starch vary depending upon the raw material. For example, maize is normally wet milled after it has been steeped initially in dilute sulfuric acid for 40–50 hours in order to soften the kernels. The milling process releases the germ which contains the oil, and the fibre is then separated from the endosperm by milling it finer. Following this, centrifuges are used to separate the starch from the protein before the starch is washed and dried.

The variation between the starch from different plants is considerable. The percentage of amylose varies from 27% in maize starch through to 22% in potato starch and 17% in tapioca starch. Waxy maize starches are unusual in that they are almost pure amylopectin. This is extremely convenient because it avoids the need to separate amylopectin from amylose chemically.

The Cooking of Starch

A fundamental difference between starch and the other gelling agents is that starch has to be cooked rather than dissolved. Indeed, raw starch is insoluble. When starch is examined under the microscope it is seen to consist of discrete granules. The shape of the granule depends upon the origin of the starch.

These granules contain micelles of starch molecules. When the granule is heated in water at a given temperature the granules swell and start to absorb water. This process is called gelatinisation, the temperature at which this happens being the gelatinisation temperature. This temperature is a characteristic of the different types of starch; for example, maize starch gelatinises from 64–72 °C. There is an exception to this and that is waxy maize starch which forms non-gelling, clear fluid pastes. Waxy maize starch behaves as a gum rather than a gelling agent, and this is one of the types of starch that are used as a substitute for gum acacia.

The changes that occur when starch is heated in water can be studied in a number of ways. One way is to follow the changes under a microscope whereas another is to measure the viscosity of the paste. The instrument normally used to monitor paste viscosity is the Brabender amylograph. Upon cooking maize starch, the viscosity increases when the starch begins to gelatinise. As the temperature rises towards 95 °C the viscosity falls. When the paste is cooled the viscosity rapidly increases. This variation of viscosity with temperature is characteristic for each different origin of starch. Potato starch, for example, has a lower gelatinisation temperature than maize starch but a higher maximum viscosity, and when cooled the viscosity of potato starch rises less. Once again, amylopectin starches do not show this behaviour as they do not gel.

Obtaining Different Properties in the Starch

There are two methods of obtaining a starch with different properties. One is the biological approach of using different types of plant – the best example of this is waxy maize which yields a starch that is nearly pure amylopectin. The other method is to modify the starch chemically – chemically modified starch is normally declared as 'modified starch'. There is a whole range of modified starches available and there is, of course, no bar to chemically modifying a starch from a special source.

Modifications that are performed on starch can be compared to the modifications that are performed on other polymers, *e.g.* the vulcanisation of rubber. Similarly, starch is treated chemically to give greater stability to heat and shear.

The Use of Starch in Confectionery

Starch has two classes of use in confectionery: gelling and non-gelling. Where the starch is non-gelling it is usually being used as a substitute for a gum. Native starch is little used in confectionery – probably the only use of native starch is where it is used to dust a sticky product, *e.g.* Turkish delight.

Thin Boiling Starches

These are the commonest type of starch used in confectionery and are used to make jellies, pastilles and wine gums. They are made by heating a dispersion of the starch with a small quantity of acid at a temperature below the gelatinisation point, where the effect is to reduce the molecular weight by hydrolysing a few of the bonds. This decreases the viscosity of the starch pastes as might be expected from theory. Starch suppliers have devised a system where these starches are classified by fluidity, where the fluidity is the reciprocal of the viscosity. Thus, in a series of fluidity numbers 20, 40, 60, 75 and 80, the 80 fluidity gives the lowest viscosity paste.

Pre-gelatinised Starches

These starches have been gelatinised either by extrusion or by heating in water followed by roller drying. These products are not normally used in confectionery although they might be of interest in extruded products.

Oxidised Starches

The effect of oxidation is to reduce the tendency to form micelles. This reduces the tendency to gel as well as making the paste more stable. The usual oxidising agent is hypochlorite.

Non-gelling Starches

These products are intended for uses where the starch replaces a gum such as gum acacia. A typical product for this use is an oxidised waxy maize starch.

Gum Tragacanth, E413

Gum tragacanth comes from the shrub *Astragalus*. Its major country of origin is Iran. The gum is normally available as yellowish pieces or as a spray dried powder. It is another polysaccharide polymer and is composed of glucuronic acid and arabinose, the molecular weight being around 840 000. It is odourless but has a mucilaginous taste.

Gum tragacanth is insoluble in alcohol but is soluble in alkaline and aqueous hydrogen peroxide solutions. When dispersed in cold water it gives a thick paste.

This material is used in small quantities in confectionery, particularly in making lozenges. It is hard to find an alternative product that works as well, although a mixture of gum acacia and gum tragacanth is sometimes used to replace pure gum tragacanth in lozenges. The mixture is used purely as an economy measure as pure gum acacia is not an acceptable substitute.

Locust Bean or Carob Bean Gum

This material is another plant polysaccharide. The source is the seeds of the carob tree *Ceratonia siliqua* also known as the locust bean tree. The trees grow around the Mediterranean and in California. An alternative name for the fruit is 'Saint John's Bread'. An impure material called carob pod flour can be produced by removing just the hulls and milling the endosperms directly. An impure product like this will give a cloudy solution in water. To produce a pure material the gum is dissolved from the seeds with hot water. The gum solution is then purified by diatomaceous earth filtration. Next, the gum is precipitated by adding propan-2-ol, and the resulting precipitate is pressed followed by washing with pure alcohol to dehydrate it. Further pressing is used to recover the alcohol and the product is then dried and milled to the required size.

Chemically, the gum is a galactomannan composed of β-D-mannose and α-D-galactose units. The β-D-mannose units are linked $(1\rightarrow4)$ to make the main chain while the α-D-galactose units form the side chain. Locust bean gum is very similar to guar gum; indeed, it was developed as a substitute when locust bean gum was not available. The two gums differ in the ratio of D-galactose to D-mannose: in guar gum the ratio is 1:2 whereas in locust bean gum the ratio is 1:4, *i.e.* the difference lies in the number of galactose residues attached to the main D-mannose chain.

Locust bean gum used on its own is a thickener – it will dissolve in water at 80 °C. When used with κ-carrageenan the substances exhibit synergy in producing an elastic and very cohesive gel. A similar synergy occurs with xanthan gum (see below), again producing an elastic and very cohesive gel. Locust bean gum is also used to stiffen agar jellies although, in general, this gum is too viscous on its own to be much used in confectionery.

Xanthan Gum

This gum was the first microbial gum to be used in confectionery. It is produced by the aerobic fermentation of *Xanthomonas campestris*, where a specially selected culture is grown on a carbohydrate-containing nutrient medium with a nitrogen source and other essential elements. The pH, temperature and aeration are controlled carefully, the resulting product is sterilised and the gum is precipitated upon addition of propan-2-ol. After washing, the precipitate is pressed to remove residual alcohol before being dried and ground to the required size.

Chemically, xanthan gum is an anionic polysaccharide with monomers of D-glucose, D-mannose and D-glucuronic acid. The polymer backbone is composed of $(1\rightarrow4)$-linked β-D-glucose units similar to cellulose. On alternate glucose units a trisaccharide chain containing one glucuronic acid and two mannose residues is fixed to the 3-position. This gives a stiff chain which can form single, double or triple helices. The molecular weight is approximately 2×10^6 – the molecular weight distribution is narrower than for most polysaccharides.

A solution containing 1% xanthan gum has a pH of between 6.1 and 8.1, and solutions behave as though they are a complex network of entangled rod-like molecules. Xanthan gum normally functions as a thickener but combines synergistically with locust bean gum to produce a very cohesive and elastic gel. Xanthan gum is little used in confectionery but it is one of the few substances that can be used as a substitute for gum tragacanth in lozenges.

Egg Albumen

In this work the name 'egg albumen' is used to refer to the mixture of proteins in egg white. Around 54% of the protein in egg albumen is ovalbumen. With improvement of analytical techniques the number of proteins identified in egg albumen will continue to increase, but unless fractionated egg albumen proteins become available this will be of little consequence for confectionery makers.

Practical Forms of Egg Albumen

Fresh egg white is not normally used in confectionery as it is too unstable and could have bacteriological problems. In practice, various forms of dried albumen are used – these have the advantage that they can be thoroughly tested bacteriologically before use. The form most commonly used in confectionery is dried egg white which is typically made by pouring egg white into shallow trays and drying it. The resulting sheets are then ground to final size. This type of product is a low technology product although it is possible to apply sophisticated drying methods, *e.g.* spray drying, to produce egg albumen that will reconstitute to a product similar to fresh egg white. 'Fluff dried' albumen is made by whipping the albumen followed by rapid drying of the resulting foam.

Although nougat must originally have been made from fresh egg white, dried albumen works just as well. The sophisticated forms of egg albumen, however, do give superior results in some bakery applications. The anecdote is often repeated, particularly by marketing men, of the cake mix that sold better when the instructions were changed to 'add an egg'. In practice, a cake mix formulated from old-fashioned dried egg albumen produces a rather unsatisfactory cake. In this application one of the more sophisticated egg albumen products is needed but is worth the extra cost.

Properties of Egg Albumen

Egg albumen is normally used in confectionery for two reasons: it whips into a foam, and the foam can be set irreversibly by heat. An advantage of egg albumen is that, unlike some other whipping agents, it is relatively unaffected by the presence of fat, which usually acts as a foam breaker in these systems.

In confectionery systems egg albumen is usually set by beating the reconstituted egg albumen into a hot sugar syrup. The coagulation temperature is affected by the water activity. For example, in a 40% sugar syrup the coagulation temperature is 65 °C but in a 60% sugar syrup the coagulation temperature rises to 75 °C. Another advantage of egg albumen is that it always coagulates reliably.

Testing Egg Albumen

Egg albumen needs to be tested thoroughly: it needs to be tested chemically to see that it is not contaminated with heavy metals (as do all food materials); it needs to be rigorously tested for microbiological contamination – this is particularly important as there is a risk of *Salmonella* contamination. Egg albumen also, unlike some other materials, needs to be tested to see that it will actually perform satisfactorily in the product in question, *e.g.* a sample of egg albumen that is entirely satisfactory for making nougat might be unsatisfactory in a cake mix. This can obviously be done by making an experimental batch of product. However, often this is not convenient, particularly where making a small batch is not easy or where the production time is long. This problem has been tackled by devising various empirical tests. In theory, an empirical test should work even if the theory of its operation is not understood, although it has happened that empirical tests have been found, upon examination, to be of low predictive value.

As an example, a typical empirical test might be to mix a given weight of egg albumen in a given volume of water in a particular mixer for a specified time. The resulting foam is poured into a funnel with a weighed tube under the spout and the height of the foam measured initially and after a length of time. The weight of liquid that runs into the tube is then measured. The results of this sort of test are then compared with the specification. The egg albumen is then rejected or accepted accordingly. In one case the test protocol specified a particular model of food mixer.

When this mixer was discontinued the testers were forced to look around for an alternative type of mixer and to recalibrate the test.

The variation in the performance of different types of egg albumen in different systems is almost certainly caused by variations in the degree of denaturation of the protein. Those products that work best with fresh egg white clearly need an undenatured product.

Substitutes for Egg Albumen

Milk Proteins. As some milk proteins will gel on heating and others can be modified to make whipping agents it has long been thought that milk proteins could be used as a whole or partial substitute for egg proteins. Purified whey proteins were regarded as a suitable raw material as whey is a low value by-product from cheese-making although the early products in this area were not very successful for a couple of reasons. One was that residual fat in the product inhibited foaming. A more serious problem was that active lipase enzymes in the ingredient were introduced into the finished product. As discussed on page 30, lipases break fats down to their constituent fatty acids, and in the case of butter fat the principal product is butyric acid, which, at higher levels imparts an unpopular cheesy flavour to confectionery. Similarly, lauric acid is the principal product of lauric fat, a component of nut oils. As soap normally contains sodium laurate it is not too surprising that free lauric acid tastes of soap. A nougat, which is the sort of product where egg albumen is used, could contain both lauric fat and nuts. Introducing lipases is generally detrimental to product shelf-life.

Satisfactory products based on milk protein have, however, been produced. One such is the range of products sold under the trade mark *Hyfoama*. This material is normally used in conjunction with other gelling or whipping agents. Typically, *Hyfoama* is used with egg albumen as the properties of the two substances complement each other – *Hyfoama* foams reliably but does not coagulate reliably; in contrast, egg albumen always coagulates reliably but sometimes foams badly. All foaming agents are sensitive to the presence of fat and *Hyfoama* is no exception.

Soya Proteins. Early attempts to make albumen substitutes from soya protein also ran into problems, for example, a bean flavour tended to appear in the finished product. A solution to these problems has been found and whipping agents based on enzyme modified soy proteins are now available. The advantage of enzymatic modification is that by appropriate choice of enzymes the protein can be modified in a very controlled way – chemical treatment is far less specific. In making these materials the manufacturer has full control of the substrate and enzyme which allows the final product to be made almost to order. The substrates used are oil-free soy flakes or flour, or soy protein concentrate or isolate. The enzymes used are chosen from a combination of pepsin, papain, ficin, trypsin or bacterial proteases. The substrate is treated with one or more enzymes under carefully controlled conditions and the finished product is then spray dried.

In use, these soya-based products, unlike egg albumen, do not coagulate. They must be used in conjunction with egg albumen or another coagulating material if coagulation is needed. The soya-based proteins have the advantage that they have approximately twice the whipping capacity of egg albumen. The modified soya protein is used by dispersing it directly in 2–3 times its own weight of water. Unlike egg albumen, it is not necessary to pre-soak this material. In a typical confectionery application such as nougat, a 50:50 mix of modified soy protein and egg albumen is used. Another advantage of these modified soy proteins is that they do not have a problem with over beating.

CHEWING GUM INGREDIENTS

Chewing gum of necessity needs an ingredient that is scientifically a rubber. This does not mean that it is made from the sort of natural rubber that household gloves and similar products are made from. The specialised materials concerned are considered below.

Chicle

Chicle is a form of natural rubber which is collected from the sap of *Achras saporata* which grows in Mexico and south and central America. The trees are tapped like rubber trees by cutting a herringbone pattern in the bark and collecting the sap. The raw gum is then boiled down to produce blocks. These blocks contain 20–30% moisture with impurities such as sand, bark and twigs. The gum is purified by multiple washing with detergent solution followed by rinsing.

Chemically, chicle is a transpolyisoprene isomer of gutta percha. When heated above 60 °C it softens. Chicle was the original gum in chewing gum. It is claimed that it was originally brought to the United States of America by General Di Santa Anna who led the Mexicans at the Alamo.

Jelutong

This is the other natural rubber used in making chewing gum. It is produced from the sap of *Dyera costulata* of the genus apocyanesas. This tree grows in the Far East. The crude product is processed by a system involving the injection of steam into the raw material.

Chapter 4

Emulsifiers, Colours and Flavours

EMULSIFIERS

An emulsion is a disperse system of two immiscible liquids, typically an aqueous phase and a lipid phase, and occurs commonly in food systems. Emulsifiers are a class of substances that help to form or stabilise an emulsion (Figure 4.1). Some natural products, particularly gums and proteins, act as emulsifiers. Such natural products often escape being defined legally as emulsifiers even though they are undoubtedly emulsifiers in practice. Substances capable of acting as emulsifiers tend to have one part of the molecule that is best suited to oily surroundings, *i.e.* it is said to be lipophilic, while the other end of the molecule is best in an aqueous environment, *i.e.* it is hydrophilic. The two opposite terms – hydrophobic (meaning water-hating) and lipophobic (meaning fat-hating) – are also in use. Emulsifiers are classified by a system of HLB numbers which refer to the ratio of hydrophilic to lipophilic groups present. Molecules with both hydrophilic and lipophilic groups are referred to as amphiphilic, and emulsifiers, whether natural or synthetic in origin, tend to be amphiphilic. An amphiphilic molecule is likely to be in its lowest energy state at the interface between an oil and water mix (Figure 4.2), although this diagram is an over simplification of the nature of the interface in a real system since real food systems tend to have a complex mixture of ingredients.

The sugar confectionery system that most commonly relies on emulsifiers is toffee. A typical toffee has a continuous phase of a high solids sugar syrup with milk proteins present. The disperse phase may be all milk fat, a mixture of vegetable fat and milk fat, or purely vegetable fat. The interface between the two phases is likely to be formed of some of the milk protein and any added emulsifier.

Another type of emulsion of interest in sugar confectionery is that of foams, and these can be regarded as a dispersion of air within a liquid. Some confectionery products such as whipped montelimars are a fat-in-sugar syrup emulsion with air whipped in.

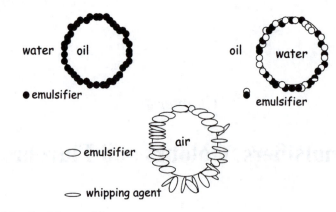

Figure 4.1   *Emulsions and foams*

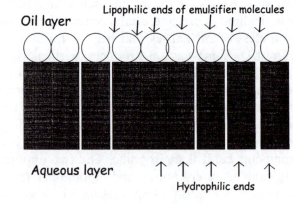

Figure 4.2   *The emulsion interface*

Sources of Emulsifiers

Some emulsifiers, *e.g.* lecithin, are purely natural products whereas others are manufactured usually from natural materials. Typical materials for manufactured emulsifiers are: vegetable oils, *e.g.* soya bean oil or palm oil; animal fats, *e.g.* lard or tallow; and glycerol. Where required, some manufacturers can supply products with kosher or halal certificates. Other raw materials are organic acids such as fatty acids, citric acid, acetic acid and tartaric acid, in addition to sorbitol and propylene glycol.

One property that often affects the performance of emulsifiers is purity. A very pure emulsifier performs very differently to the same major ingredient present at a lower purity. This is particularly apparent with monoglycerides where these are available as distilled monoglycerides, produced by molecular distillation, or in grades of lesser purity.

Legislation

Emulsifiers, naturally, tend to attract the attention of food legislators, and it is entirely reasonable that only those substances that are safe for food use are permitted. The thing that is difficult to understand is why the permitted emulsifiers vary so much between countries. The EU is, however, working towards rationalisation in this area.

Examples of Emulsifiers

Distilled Monoglycerides, E471

These are high purity monoglycerides prepared by molecular distillation.

Lecithin

As mentioned above, lecithin is a naturally-occurring emulsifier and is even believed by some to be a health food in its own right. Its discovery goes back to the nineteenth century: in 1811, L.N. Vaquelin reported the presence of organically-bound phosphorus in fat-containing extracts from brain matter, and later, in 1846, M. Gobley separated an orange-coloured sticky substance from egg yolk. This orange substance was found to have excellent emulsifying properties and Gobley named it 'lecithin', a name derived from *lekithos*, the Greek for egg yolk.

Lecithin is a polar lipid, the definition being that it is a lipid that is insoluble in acetone. It is one of a whole class of phospholipids, which tend to be found in the membranes of animals and also in plants.

The Definition of Lecithin. The definition that is used for food lecithin is 'a mixture of polar and neutral lipids with a polar lipid content of at least 60%'. Note that this is different from the scientific usage where lecithin is used as a trivial name for phosphatidylcholine.

Sources of Lecithin. The main commercial source of lecithin is the soy bean although lecithins are also produced from sunflower seed, rapeseed, maize, and, in small quantities, peanuts. (Lecithin can indeed be produced from egg yolk but this is not commercially competitive.) In the future it might be possible to produce lecithins from processes involving micro-organisms.

The Production of Soy Lecithin. The soy beans are cleaned, de-hulled and cracked, followed by rolling to obtain thin flakes. These flakes are then treated with solvent, where the resulting mixture is known as a miscella. After filtration the solvent is removed by vacuum distillation, leaving the residue, a reddish yellow oil, which contains some 2% impure lecithin. The oil and lecithin mix is heated to 70–90 °C and

Table 4.1 *Composition of oil-free soy lecithin*

Substance	% composition
Phosphatidylcholine	22
Phosphatidylethanolamine	23
Phosphatidylserine	2
Phosphatidylinositol	20
Phosphatidic acid	5
Phytoglycolipids	13
Other phospholipids	12

(Source: Lucas Meyer, Lecithin Properties and Applications)

mixed with 1–4% of water – this causes the lecithin to swell and precipitate as a jelly-like mass. This jelly-like mass is then removed by specially designed high speed separators to give the separated product, a sludge containing approximately 12% soy bean oil, 33% phospholipids and 55% water. This material is then treated in a thin film vacuum evaporator to remove almost all of the water. The resulting product has 60–70% polar lipids, 27–37% soy bean oil, 0.5–1.5 % moisture and 0.5–2% impurities. This product is the ordinary soy lecithin of commerce, the composition of which is shown in Table 4.1.

It is possible to remove the soy bean oil in order to produce a de-oiled lecithin but in confectionery use there is little point in using a de-oiled product. Chemical modification of lecithins is possible but this would cause them to lose their natural status. Another way of modifying the properties is to fractionate the raw lecithin in order to yield products that are richer in one of the components. The resulting products, of course, retain their natural status.

Sucrose Esters, E473

These emulsifiers (**1**) are prepared from sucrose and edible fatty acids. The primary hydroxyl groups of the sucrose are esterified by the fatty acid. It is possible to react fatty acids with one, two or three primary hydroxyl groups to yield mono-, di- or tri-esters, respectively.

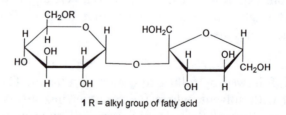

1 R = alkyl group of fatty acid

One advantage of sucrose esters is that they can be made with a wider range of HLB values than other emulsifiers. In Figure 4.3 the HLB range

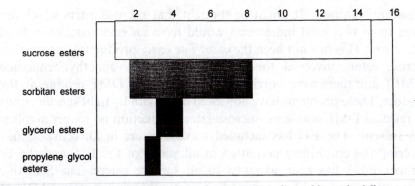

Figure 4.3 *Comparison of the range of HLB value obtainable with different ester emulsifiers*

covered by sucrose esters is compared with the HLB range obtainable with sorbitan esters, glycerol esters and propylene glycol esters. These materials are all families of emulsifiers that are chemically esters. As an emulsifier needs to be amphiphilic, esters are a popular structure for synthetic emulsifiers. Available grades of sucrose esters cover the range of HLB from 2 to 15, and this wide range of values is obtained by varying the monoester content from 10 to 70% (Table 4.2). Thus, a high HLB emulsifier is suitable for use in an oil-in-water emulsion whereas a low HLB emulsifier is used in a water-in-oil emulsion.

The practical effect of this very wide HLB range is that sucrose esters can be used in a very wide range of confectionery products (Table 4.3). It should, however, be remembered that it is not necessarily the same sucrose ester used in each type of product. Another useful property of sucrose esters is that they are stable up to 180 °C.

Table 4.2 *Effect of monoester content on HLB for sucrose esters*

Monoester content (%)	HLB value
70	15
50	11
30	6
10	2

Table 4.3 *Uses of sucrose esters in confectionery*

Product	Use level (%)
Soft chewy confectionery	0.1–2.0
Chewing gum	0.2–0.4
Tabletted products	0.1–2.0
Caramels and toffees	0.1–0.5

Regulatory Status. It might be thought that sucrose esters which are made from two food ingredients would have an easy passage in food legislation. This has not been the case. The early production method for sucrose esters involved the use of the solvent dimethylformamide (DMF), and there were worries about the effect of DMF residues in the product. These problems have now been dealt with by tight specifications on residual DMF and some sucrose ester production no longer involves this solvent. The EU has included sucrose esters in Directive 78/663 covering the emulsifiers permitted in all states of the EU, and the E-number E473 has been assigned to all sucrose esters. The Scientific Committee for Food has assigned an acceptable daily intake (ADI) of 20 mg per kg of body weight.

Uses of Emulsifiers in Sugar Confectionery

Emulsifiers are used in a range of sugar confectionery products although they are not normally used in fat-free products such as boiled sweets, gums or jellies. The usual use of emulsifiers in sugar confectionery is to keep oils or fats dispersed. A subsidiary effect is also to alter the texture since the texture of a product is affected by the size of any dispersed fat globules, and a related effect to this is the handling of the product during manufacture since the product will need to flow and probably be shaped and cut. Adding the wrong emulsifier or an excess of emulsifier can cause handling problems. If a product as made has a film of oil on the surface, it acts as a lubricant and the product flows easily. If a highly effective emulsifier is used the oily film will not form and the product will stick to conveyor belts and cutting knives. In some cases polytetrafluoroethylene (PTFE) coated conveyors and cutters have had to be installed.

In this section, emulsifier means a material that is on the list of permitted emulsifiers and stabilisers – there are some materials that are extremely effective emulsifiers but which, in fact, do not qualify as emulsifiers in food law. The main examples are the proteins contained in milk: they are effective at emulsifying oils and fats but do not qualify for the legal description of emulsifiers.

Caramels and Toffees

In these products the emulsifiers are used to assist in dispersing the fat. Toffees or caramels can be made without adding any emulsifier but only at the expense of using a larger quantity of skim milk solids in the recipe. Particularly in the European Union (EU) where the Common Agricultural Policy (CAP) has increased the price of milk, reducing the milk content and emulsifiers is very common. The overall effect of the high prices is to reduce the consumption of milk and it is now rare to find a manufactured toffee or caramel that does not contain added emulsifiers. Typical emulsifiers used in toffees are distilled monoglycerides or a mixture of mono- and diglycerides, lecithin or possibly sucrose esters. Of these emulsifiers lecithin is almost universally allowed whereas sucrose esters are the most tightly restricted. In the case of the monoglycerides the performance of the material is affected by the purity of the material; distilled monoglycerides are an example of a high purity emulsifier.

Chewy Confectionery

These products are usually made by dispersing some fat in a mixture of sugar and glucose, the resulting product being flavoured with a fruit or possibly mint flavour. Some of these products contain egg albumen or other proteins. This type of confectionery normally contains an emulsifier to assist with fat dispersion and to give the desired texture.

Chewing Gum

In chewing gum, emulsifiers can act as a plasticiser. This is not too surprising as chewing gum is effectively a chewable polymer product.

Tabletted Products

In these systems, emulsifiers can be used to disperse oily ingredients such as flavours. Another use is to make it easier for the granules to flow during processing.

COLOURS

When synthetic colours were first added to confectionery the dyes used were merely batches of the sort of dye used in the textile industry. Now, of course, the use of colours in foods is strictly regulated and food colours are rigorously tested to ensure that they are not harmful. Governments around the world have lists of permitted colours – unfortunately, these lists of permitted colours are different throughout the world. It might be thought that some scientific consistency could be achieved but this is not the case. Indeed, some manufacturers who produce products for the international export markets are reduced to leaving out all of the colours as a way of making them universally acceptable.

In taste trials, colour has been found to have an important influence upon flavour perception. Early fruit-flavoured products were probably flavoured with jam and did not have a particularly strong flavour. Even with modern flavours the experiment of putting the 'wrong' colour in the product does cause an appreciable proportion of tasters to misidentify the flavour.

Most gum or jelly products if made uncoloured will have a yellow or orange colour. This might pass for an orange- or lemon-flavoured sweet but would not pass otherwise. An approach that is used when making uncoloured products is to wrap uncoloured sweets in a coloured wrapper that represents the flavour concerned.

Technical Requirements of Colours in Sugar Confectionery

To be used successfully in sugar confectionery a food colour needs the following attributes as well as complying with the appropriate legislation: it should be stable to heat and light; it should be stable to reducing sugars; and resistance to sulfur dioxide is also useful. Most of the colours used in sugar confectionery are water soluble. This is simply convenient as most sugar confectionery products contain very little fat anyway.

Synthetic Colours

Synthetic colours are available for almost all possible shades (Table 4.4), where the intermediate shades can be produced by blending colours. In

Table 4.4 *Synthetic colours*[a]

Shade	Name	E-number (EU)	FD&C number (USA)	Chemistry
Red	Allura red AC	E129	Red 40	Monoazo
Red	Ponceau 4R	E124	Not permitted	Monoazo
Red	Carmoisine	E122	Not permitted	Monoazo
Red	Amaranth	E123	Not permitted	Monoazo
Red	Erythrosine BS	E127	Red 3	Xanthene
Red	Red 2G	E128	Not permitted	Monoazo
Orange or yellow	Tartrazine	E102	Yellow 5	Pyrazolone
Orange or yellow	Yellow 2G	E107	Not permitted	Monoazo
Orange or yellow	Sunset Yellow FCF	E110	Yellow 6	Monoazo
Orange or yellow	Quinoline Yellow	E104	Not permitted	Quinoline
Green	Green S (Brilliant Green BS)	E142	Not permitted	Triarylmethane
Green	Fast Green FCF	—	Green 3	Triarylmethane
Blue	Indigo Carmine	E132	Blue 2	Indigoid
Blue	Patent Blue V	E131	Not permitted	Triarylmethane
Blue	Brilliant Blue FCF	E133	Blue 1	Triarylmethane
Brown	Brown FK	E154	Not permitted	Mix[b]
Brown	Chocolate Brown FB		Not permitted	Monoazo
Brown	Chocolate Brown HT	E155	Not permitted	Diazo
Black	Black PN	E151	Not permitted	Diazo

[a] Some colours that are legal in some countries are illegal in others. This table does not attempt to be definitive. [b] Brown FK is a mixture of a monoazo and a diazo compound.

general, synthetic colours are much more stable than natural colours to light, heat and extremes of pH.

Synthetic colours can be supplied as water soluble powders, prepared solutions, easily dispersed granules, pastes or gelatine sticks. Blocks of colour in vegetable fat are also available for use in fat-based products. The attraction of soluble powders is that they are the least expensive and can be made up as required for use although the other forms have the advantage that they are at a concentration that is ready to use – the disadvantage with the pre-prepared versions is, however, usually financial.

Synthetic colours are normally so intense that they must be diluted considerably for them to be readily measured and dispersed into the product. If colour solutions are made up in the factory they have to be made up not more than 24 hours before use to avoid mould spoilage. The pre-prepared colour solutions have a permitted preservative or are made up in glycerine, propylene glycol or propan-2-ol. These non-aqueous solvents inhibit mould growth.

Lake Colours

Where an opaque colour is required a lake colour is often used. Types of sugar confectionery that fall in to this definition are panned goods, chewing gum, bubble gum, chews, fondant and toffee.

Lake colours are made by precipitating a water soluble colour with an aluminium, calcium or magnesium salt on to aluminium hydroxide. The dye content can be varied from 10 to 40%. The particle size of the finished product must be small enough to give a speck-free but economical coverage.

Lake colours have a number of advantages over the same dye in solution. They do not migrate as much and the dye is more stable to light. Migration occurs because between pH 3.5 and 9 the lake colour is insoluble. A type of product where a non-migrating dye would be useful is a striped confection where dye migration would cause the stripes to merge. Lake colours can also be used in fat-based products where the lake is dispersed in the fat.

Interference Colours

One recent innovation was the introduction of interference colours which are used to add a glow to the product. In sugar confectionery they are used on panned goods, and elsewhere have been very successful in the cosmetic business. Interference pigments have their properties because they cause a selective reflection of light where the wavelength, and hence the colour, of the reflected light depend upon the thickness of the layers of pigment. For example, if the mineral mica (potassium aluminium silicate) is coated with titanium dioxide the TiO_2 pigment particles change in apparent colour with increasing layer depth from white to cream to a golden yellow colour, then to a copper-red followed by lilac. These colour changes are produced by the thin platelets causing an interference effect. Further coating produces a shining blue followed by green. If the process is continued yet further the colours reappear in the same order.

The apparent colour of an interference pigment varies depending upon the angle from which it is viewed (see Figure 4.4). This is called the 'colour flop' or 'flip flop' effect. As an example, if a blue pigment is laid on a white background, white light is reflected at the glancing angle as blue, and only at this angle will the product appear blue. The rest of the light passes through the transparent pigment and the product appears as the complementary colour to blue which is yellow. (The yellow colour has been produced by removing the blue light from white light.) This means that from all angles except the glancing angle the product will appear to be a pale yellow, but when turned through the glancing angle it will flash blue. Products such as these are made from the mineral mica with titanium dioxide or iron oxide added – light colours contain titanium dioxide (E171) whereas dark, rich colours contain iron oxide (E172). Both types use mica as a base: a titanium dioxide coating gives coloured highlights available in gold, red, blue and green; the iron oxide imparts the deep, rich colours, available as a red brown lustre, red brown glitter, copper lustre, copper glitter, bronze lustre, bronze glitter, dark red lustre and beige lustre. These products are covered under the European guidelines on potassium aluminium silicate (mica), which has the E-number E555.

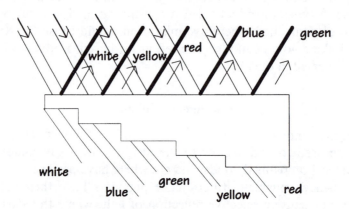

Figure 4.4   *Interference colours – increasing the thickness of the TiO_2 layer*

Natural Colours

There is a belief that natural products are inherently safer and healthier than man-made ones. This belief is lacking in intellectual rigour. Of the most toxic substances known to man the majority are natural – examples are aflatoxin, a mould metabolite, and ricin, found in castor oil beans. However, the presence of natural colours is a marketing advantage so they are used. Natural colours in general are less heat stable, less light stable, and give a less intense and less pure colour than synthetic colours. Natural colours have been used in the form of impure extracts rather than pure products. In this form higher doses are needed than with synthetic products. When purified some natural pigments are more intense in colour and can be used in lower doses than synthetic colours. One other problem with natural colours is that the range of colours available is restricted, as is seen in comparing Table 4.5 with Table 4.4 on page 66.

Table 4.5 *Natural colours*

Shade	Name	E-number	Pigment	Source
Red	Anthocyanins	E163	Anthocyanins	Grape skins, red cabbage
Red	Betanin, vulgaxanthin	E162	Betanin, vulgaxanthin	Beetroot (*Beta vulgaris*)
Red	Cochineal	E120	Carmine and carminic acid	Cochineal beetle
Yellow to red	β-Carotene	E160a	Carotene	Carrots, alfalfa
Yellow to red	Annatto	E106b	Bixin, norbixin	*Bixa orellana* seeds
Yellow to red	Lutein	E161b	Carotenoid	Tagetes or Aztec marigolds
Yellow to red	Crocin		Carotenoid	Saffron crocus, *Gardenia jasminoides*, fruit
Yellow to red	Paprika	E160c	Capsanthin and capsorubin	Sweet red pepper (*Capsicum annum*)
Yellow to red	β-Apo-8'-carotenal	E160c	β-Apo-8'-carotenal	Normally nature-identical
Yellow to red	Canthaxanthin	E161g	Canthaxanthin	Normally nature-identical
Yellow	Turmeric	E100	Curcimin	Turmeric root (*Curcuma longa*)
Yellow	Riboflavin (vitamin B$_2$)	E101	Riboflavin	Milk, yeast, normally nature-identical
Yellow	Riboflavin 5'-phosphate	E101a	Riboflavin 5'-phosphate	Normally nature-identical
Green	Chlorophyll	E140	Chlorophyll and chlorophyllins	Green leaves, alfalfa, grass
Green	Copper chlorophyll	E141	Copper chlorophyll	Derived from chlorophyll
Brown	Caramel	E150	Melanoidins	Carbohydrates heated, possibly, with ammonia
Black	Carbon black	E153	Carbon	Carbonised vegetable matter

Sources of Natural Colours

Caramel, E150

Caramel in this context means a brown colour that is produced either traditionally by heating sugar or as a very intense product that is made by heating carbohydrate, usually glucose syrup, with ammonia. Caramel colour is the product of the Maillard reaction, *i.e.* the reaction of a reducing sugar with an amino group. Chemically the colour is a melanoidin – these substances are extremely stable and can be used in any type of confectionery.

Chlorophyll, E140

This is the green pigment that is responsible for light absorption to provide energy for photosynthesis. It is widely distributed in nature: sources are green leaves, grass, alfalfa and nettles. Chlorophyll preparations are available for colouring boiled sweets, and the extract that is used is a mixture of chlorophyll with lutein and other carotenoids, which gives an olive green colour. Chlorophyll is most stable in neutral or alkaline conditions but has a limited stability to heat and light.

Copper Chlorophyll, E141

This material is made from chlorophyll but it is more blue than natural chlorophyll. The chemical modification makes it much more stable to heat and light, and hence a more useful material than natural chlorophyll.

Cochineal, E120

Cochineal is a traditional natural colour made from a Mexican beetle, *Dactylopius coccus*. The only problems with cochineal, apart from expense, are that it is not kosher and it is not animal-free. (Cochineal is not kosher, not because it is made from an insect but because the insect is itself not kosher.)

Riboflavin, E101

This is vitamin B_2. It can be extracted from yeast but is normally encountered as a nature-identical substance. Riboflavin produces an orange yellow colour. It is stable to acid but unstable in water. Unfortunately, riboflavin has an intensely bitter taste but it is sometimes used for panned goods.

Riboflavin-5-phosphate, E101a

This material is less bitter than riboflavin; its water stability is also greater than that of the unmodified material. It is normally only encountered as a pure synthetic substance. Like riboflavin, it is used on panned products.

Carbon Black, E153

This material is carbonised vegetable matter, *i.e.* very finely divided charcoal. Inevitably it is the most light-fast of all colours. Obviously, it is only available as a solid and a common use is in liquorice products.

Curcumin, E100

Curcumin is obtained from the spice turmeric which comes from the plant *Curcuma longa*, which itself is related to ginger. Curcumin is a bright yellow pigment which is obtained by extraction from the plant to give a deodorised product. It is oil soluble but is sometimes produced in a water dispersible form.

The colour of curcumin varies with the pH of the medium. Under acid conditions a bright yellow colour is obtained, but under alkaline conditions it imparts a reddish brown hue. This colour shift occurs because curcumin undergoes keto–enol tautomerism.

The most serious problem with curcumin is its instability to light. One recommendation is that curcumin should not be used in products that are exposed to light unless the moisture content is very low. A confectionery product that fits this description is boiled sweets. The heat stability of curcumin is sufficiently good that it can withstand 140 °C for 15 minutes in a boiled sweet mass.

The other stability problem with curcumin is sulfur dioxide: if the SO_2 level is above 100 ppm then the colour will fade.

Within the restrictions outlined curcumin is a successful natural colour.

Crocin

Crocin is found in saffron and gardenias. The commercial source of crocin is the gardenia bush; extracting crocin from saffron is not economically viable. Saffron is obtained from the *Crocus sativus*, and 70 000 plants are needed to produce 500 g of saffron, which contains approximately 70 g of crocin. The town of Saffron Walden in Essex takes its name because saffron used to be produced there.

Chemically, crocin is the digentiobioside of crocetin. It is one of the few water soluble carotenoids, producing a bright yellow shade in water. Unfortunately, crocin is bleached by SO_2 present at levels above 50 ppm. The heat stability of crocin is good enough to use it in boiled sweets.

Carotenoids

The carotenoids are a wide range of substances which are extremely abundant in Nature. It has been estimated that they are produced naturally at a rate of 3.5 tonnes per second, and some 400 carotenoids have been identified to date. They are found in fruits, vegetables, eggs, poultry, shellfish and spices – for example, orange juice and peel have been found to contain 24 different carotenoids.

Several carotenoids, *e.g.* β-carotene, are important as pro-vitamins – β-carotene is sometimes erroneously referred to as a vitamin when it is, in fact, pro-vitamin A. (Pro-vitamins are inactive precursors of vitamins which are synthesised by the body.) A common piece of dietary advice is to eat more carotenoids – a typical diet contains large quantities of carotenoids, much greater than any quantity that might be used as a colouring agent. The body has the ability to store fat-soluble vitamins (such as A and D) but cannot store water-soluble ones (such as vitamin C and the B vitamins), and since it cannot readily dispose of any excess, overdoses of fat-soluble vitamins can be very serious. In order to prevent vitaminosis occurring, the human body has a regulatory system that turns off the conversion of β-carotene to retinol (vitamin A) if stocks are adequate. Thus, using carotenoids as food colours does not pose a risk of vitamin A overdose.

Legally, carotenoids are divided between two E-numbers: E160 covers the carotenoid hydrocarbons β-carotene, lycopene and paprika, as well as the apo-carotenoids, *e.g.* bixin; E161 covers the xanthophylls and the carotenoids lutein, astaxanthin and canthaxanthin.

Most carotenoids are fat-soluble although, like curcumin, preparations that allow them to be dispersed in water are made. The colours available from carotenoids vary between pale yellow and red.

Chemically, carotenoids have conjugated double bonds which render them liable to oxidation. However, this tendency to oxidation can be reduced by adding antioxidants to the product. In the sort of product where natural colours are used, suitable antioxidants are tocopherols or ascorbic acid. Chemical antioxidants such as butylated hydroxytoluene might be suitable technically but would not fit the image of an all natural product. To create a more favourable image, ascorbic acid could be declared as vitamin C rather than as an antioxidant. As oxidation can be started by exposure to light, unsuitable storage conditions are best avoided.

Carotenoids are generally stable to heat. The levels required can be as low as 10 ppm. In the case of β-carotene this material is available as a nature-identical form.

β-Carotene, E160(a)

The EU classifies β-carotene as E160(a), and natural sources that are exploited commercially are carrots and algae. β-Carotene is an oil soluble pigment although forms which can be dispersed in water are available. It is stable to heat, SO_2 and pH changes; it is, however, sensitive to oxidation, particularly when exposed to light. Depending upon the concentration the colour obtained varies between yellow and orange and it is used successfully to colour boiled sweets and other confectionery products.

Annatto, E160(b)

Annatto is classified as E160(b). It is extracted from the seeds of the tree *Bixa orellana* which grows in America, India and East Africa. The extract is a mixture of two pigments, bixin and nor-bixin. Bixin is oil soluble, whereas nor-bixin is water soluble (it is one of the two water soluble carotenes – the other is crocin), and both of the pigments are diapo-carotenoids. Annatto has long been used as a food colouring and also has some uses as a food flavouring.

Both bixin and nor-bixin produce orange-coloured solutions, and consistent with their solubilities mentioned above, bixin produces an orange solution in oily media whereas nor-bixin produces an orange aqueous solution. Obviously, bixin is the product of choice for high fat systems and nor-bixin for aqueous systems.

Nor-bixin is damaged by SO_2 if the concentration exceeds 100 ppm, and acidic conditions or divalent cations, particularly calcium, can cause nor-bixin to precipitate. These problems are tackled by producing nor-bixin preparations with buffers and sequestrants.

Nor-bixin is relatively stable to heat. The most severe conditions either isomerise the pigment or shorten the chain: either of these changes make the pigment more yellow. Nor-bixin can associate with protein which stabilises the nor-bixin – the other effect of this association is to make the colour redder.

Lutein, E161(b)

Lutein is one of the four most common carotenoids found in Nature. The EU classifies it as E161(b). Although lutein occurs in all green leafy vegetation, egg yolks and in some flowers, the commercial sources are petals of the Aztec marigold and to a lesser extent alfalfa. Purified alfalfa gives a clean, bright lemon yellow shade. Chemically, lutein is a xanthophyll and hence chemically similar to β-carotene. It is more stable to oxidation than the other carotenoids and is also resistant to the action of SO_2. Lutein is oil soluble and is most effective dissolved in oil although aqueous dispersible preparations based on lutein are available.

Betalaines

Betanin is the main pigment in the concentrated colour beet red. Approximately 80% of the pigment present in beetroot is betanin, and the pure pigment is obtained by aqueous extraction of the red table beet.

In an aqueous solution betanin gives a bright bluish red. The pure pigment is so intensely coloured that dose levels of a few parts per million are satisfactory. However, the problems with betanin relate to stability. Betanin is extremely sensitive to prolonged heat treatment although short spells such as ultra high temperature (UHT) are tolerated. Other conditions that make betanin unstable are oxygen, SO_2 and high water activity but as confectionery is a low water activity system without SO_2 or oxygen, betanin can be used.

Anthocyanins

Anthocyanins are water soluble and are responsible for the colour of most red fruits and berries. Some 200 individual anthocyanins have been identified. It has been estimated that consumption of anthocyanins is, on average, 200 mg per day, which is several times greater than the average consumption of colouring material. There are claims made that consuming anthocyanins has health benefits.

Chemically, anthocyanins are glycosides of anthocyanidins and are based on a 2-phenylbenzopyrilium structure. The properties of the anthocyanins depend on the anthocyanidins from which they originate.

The anthocyanins are extracted commercially using either acidified water or alcohol. The extract is vacuum evaporated in order to produce a commercial colour concentrate. The raw materials can be blackcurrants, hibiscus, elderberry, red cabbage or black grape skins, the most commonly used commercially being black grape skins which can be obtained as a by-product from grape-juice or wine making.

The colour normally given by anthocyanins is a purplish red. Anthocyanins are amphoteric, and of the four major pH-dependent forms the most important are the red flavylium cation and the blue quinodial base. At pHs up to 3.8, commercial anthocyanin colours are ruby red; as the pH becomes less acid the colour shifts to blue. With increasing pH the colour also becomes less intense and the anthocyanin becomes less stable. Hence, the usual recommendation is that anthocyanins should only be used where the pH of the product is below 4.2. As these colours are considered for use in fruit-flavoured confectionery this is not too much of a problem. Anthocyanins are also sufficiently heat resistant that they do not have a problem in confectionery – colour loss and browning is only a problem if the product is held at elevated temperatures for a long while. Sulfur dioxide can bleach anthocyanins, with the monomeric anthocyanins being the most susceptible, although anthocyanins that are polymeric or condensed with other flavonoids are more resistant. The reaction with SO_2 is, however, reversible.

Anthocyanins can form complexes with metal ions such as tin, iron and aluminium, where formation of a complex normally alters the colour from red to blue. Complex formation can be minimised by adding a chelating agent such as the citrate ion.

Anthocyanins can also form complexes with proteins, a problem which can lead to precipitation in extreme cases. This problem is normally minimised by careful selection of the anthocyanin.

FLAVOURS

Chemically, flavours are complex substances,[1] and it is convenient to divide them into three groups: natural, nature-identical and synthetic.

Natural Flavours

These can be the natural material itself, *e.g.* pieces of vanilla pod, or an extract, *i.e.* vanilla essence. Extracts can be prepared in a number of ways: one is to distil or to steam distil the material of interest; another is to extract the raw material with a solvent, *e.g.* ethanol. Alternatively, some materials are extracted by coating the leaves of a plant with cocoa butter and allowing the material of interest to migrate into the cocoa butter. These techniques are also used in preparing perfumery ingredients; indeed, materials like orange oil are used in both flavours and perfumes.

In practice, some natural flavours work very well and any problems with them are financial rather than technical. Examples of satisfactory natural flavours are vanilla or any of the citrus fruits. Some other flavours, however, are never very satisfactory when all natural. (It should be noted that citrus oils are prepared from the skin rather than the fruit.) Some gum and jelly sweets contain fruit juice or pulp in addition to the fruit flavour. If a product is made with both fruit juice and fruit flavour, the experiment of leaving out the juice and adding the flavour, or leaving out the flavour and adding the juice, shows that the product with the flavour but without the juice tastes strongly of the fruit whereas with the juice alone it gives very little taste.

The Image of Natural Products

The view exists that natural products are inherently safer and more healthy than synthetic materials. There is a legal distinction applied between an ingredient and an additive. In the UK, additives generally need approval for their use to be allowed in foods whereas natural ingredients, provided that their use is traditional, do not. This system does not always appear to be ideal, as periodically some natural substance is tested and found to have a potential risk that was previously unknown.

Nature-identical Flavourings

These are ingredients that are identical to those that occur in Nature, but which are synthetically derived. From time to time it emerges that one substance produces a given flavour; for example, most chemists know that benzaldehyde has a smell of almonds. If a natural flavouring can be represented by a single substance, and if that substance can be synthesised, then the flavour is likely to be available as a nature-identical flavour. Vanilla flavour is a good example. Vanilla flavour can be all natural and derived from vanilla pods, nature-identical or artificial. The nature-identical product is based on vanillin, which is present in vanilla pods and has a flavour of vanilla. The artificial vanilla flavour will most likely be ethyl vanillin, which is not present in vanilla pods but has a flavour two and a half times stronger on a weight basis than vanillin. The claim 'nature-identical' does not seem to be much appreciated in the English speaking countries whereas in some other countries it is an important claim for marketing purposes.

The qualification for nature-identical varies between jurisdictions. In the EU, ethyl acetate made from fermented ethyl alcohol and fermented acetic acid is nature-identical. In the USA, provided that the ethyl alcohol and acetic acid are natural, *i.e.* produced by fermentation, then the ethyl acetate is classed as natural.

Practical flavours often contain a mixture of substances: some natural, some nature-identical, some synthetic. UK law classifies a flavour that contains any nature-identical components as nature-identical even though the rest of the flavour is natural. Similarly, the presence of any artificial components renders the flavour artificial.

The Case of Nature-identical Flavours

Nature-identical claims are not much appreciated in English speaking countries, but in German speaking countries the claim is more popular. Initially, it is difficult to see why a synthetic substance that happens to be present in Nature should be preferred over a synthetic substance that is not found as a naturally-occurring substance. Presumably the advantage of a nature-identical substance is assumed because it is assumed to be inherently safe. This is a paradox since synthetic substances are normally tested for safety much more exhaustively than natural ones. However, nature-identical flavours do have the advantage over natural products in that the price or quality are not affected by adverse harvests.

Synthetic Flavours

These are flavours that are produced synthetically but which are not present in a natural material. Synthetic flavours are made from a mix of flavouring substances that have been found to produce a given flavour 'note', and those who develop flavours are referred to as flavourists. Flavourists take the musical analogy of notes further by referring to the top- and bottom-notes of a flavour.

Flavour research is driven by a need to find compounds that produce desirable flavours. In some cases the improvement that is sought over a natural substance is not flavour intensity or cheapness but chemical stability.

One view of the way that flavours work is that they interact with certain receptors in the nose. Any other compound that has the same shape will work as well.

A typical synthetic flavour is a very complex mixture of substances. The mixture used will have been chosen to give the desired properties in the system of choice. Compounding flavours is a mixture of chemistry and sensory skills – flavourists spend many years learning how to produce flavours.

Dosing

Whether the flavour used is natural, nature-identical, synthetic or a mixture it has to be dosed into the product. Although some flavourings are very intense the volume added to the product has to be large enough for the equipment, or the people, to add it with sufficient accuracy, and the flavour, of course, has to be uniformly distributed within the product. This normally means producing the flavour as a solution. Flavours used in certain products are prepared with the manufacturing conditions in mind. As an example, citrus oil-based flavours can be dissolved in various alcohols. A flavour intended for boiled sweets where the flavour is likely to be added to a hot mass is unlikely to be dissolved in ethanol as the flavour should be added at a temperature around the solvent's boiling point.

Developments in Flavours

The application of ever-improving analytical methods continues to discover new flavouring compounds, be they natural, nature-identical or synthetic. The new science of chemometrics has developed to cope with the situation where chromatograms with hundreds of compounds are obtained.

Biotechnology can also be applied to produce flavouring substances. For example, if the gene responsible for producing a given substance can be identified, then in theory that gene can be expressed in other organisms. No doubt the legislators will examine whether such products qualify as natural or nature-identical and will come to a number of different conclusions, although conventional plant breeding methods have an established use in producing varieties of aromatic plants that give flavours with improved characteristics.

It remains an interesting speculation as to what would happen if a mutation of vanilla was produced that gave ethyl vanillin rather than vanillin. The new variety would be much more potent as a flavour. However, ethyl vanillin might then have to be classified as nature-identical.

Antioxidants

Antioxidants retard the oxidative rancidity of fats. Oxidative rancidity comes on suddenly rather than gradually and the problem is caused by the addition of oxygen free-radicals across any double bonds present. For example, with epoxide compounds, rupture of the initial epoxide leads to the production of various aldehydes and ketones – these can be very odiferous. One problem with oxidative rancidity is that it is a zero free energy (ΔG) process and is therefore not retarded by lowering the temperature.

Antioxidants work by being a free-radical trap, *i.e.* they readily combine with oxygen free-radicals to produce stable compounds. Any compound which has this ability is a potential antioxidant. Of course, not all such compounds are suitable for use in foods: to be used a compound has to be non-toxic and also must have legal approval. Some antioxidants are synthetic, a few are natural and a few are nature-identical.

Synthetic Antioxidants

The commonest synthetic antioxidants are butylated hydroxyanisole (BHA) and butylated hydroxytoluene (BHT). Other synthetic anti-oxidants are *n*-propyl gallate and *n*-octyl gallate.

Tocopherols

Tocopherols (**2**) are the major class of natural or nature-identical antioxidants. They occur naturally in many plant tissues, particularly

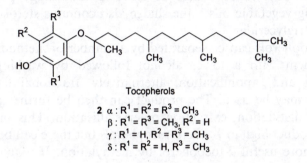

Tocopherols

2 α : R^1 = R^2 = R^3 = CH$_3$
β : R^1 = R^3 = CH$_3$, R^2 = H
γ : R^1 = H, R^2 = R^3 = CH$_3$
δ : R^1 = R^2 = H, R^3 = CH$_3$

vegetable oils, nuts, fruits and vegetables – wheat germ, maize, sunflower seeds, rapeseed, soybean oil, alfalfa and lettuce are all rich sources of tocopherols. Chemically, the structure is a 6-chromanol ring with a phytol side chain. α-, β-, γ- and δ-Tocopherols differ only in the number of methyl groups on the aromatic ring. At ambient temperatures the antioxidant activity is in the order $\alpha > \beta > \gamma > \delta$. At higher tempera-tures (50–100 °C) the order is reversed to give $\delta > \gamma > \beta > \alpha$. α-Toco-pherol acetate (**3**) is not an antioxidant since the active hydroxyl group is protected. The interest in this substance arises because under appropriate conditions, *e.g.* aqueous acidic systems, the tocopherol acetate is slowly hydrolysed to give tocopherol.

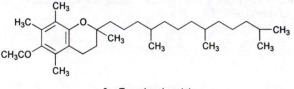

3 α-Tocopherol acetate

Tocopherols are pale yellow, viscous, oily substances which are insoluble in water but soluble in fats and oils. α-Tocopherol and its acetate are made synthetically, the synthetic products being racemates and designated DL-α-tocopherol and similarly DL-α-tocopherol acetate. These are mixtures of the four racemates.

As antioxidants, tocopherols are not as effective as the synthetic antioxidants such as BHA or BHT. However, the antioxidant effect of tocopherols is increased by mixing them with ascorbyl palmitate, ascorbic acid, lecithin or citric acid. Typical confectionery applications for these antioxidants are use with ascorbyl palmitate, lecithin or citric acid in the fat phase of toffees or caramels. Chewing gum base can also be treated with α- and γ-tocopherol to extend its shelf life.

Natural Tocopherols, E306.   The antioxidant E306 is defined as 'the extract of natural origin rich in tocopherols'. It is referred to as natural vitamin E. The major source of E306 is the sludge produced by deodorising vegetable oils – the sludge also contains sterols, free fatty acids and triglycerides.

The tocopherols can be separated by a number of methods: one is to esterify them with a lower alcohol followed by washing, vacuum distillation and saponification; alternatively, fractional liquid–liquid extraction may be used. The product can then be further purified by molecular distillation, extraction or crystallisation. This process produces a product high in γ- and δ-tocopherols but these can be converted into the more useful α-tocopherol by methylation. If required α-tocopherol acetate can be made by acetylating the α-tocopherol.

α-Tocopherol, E307.   This material is also known as synthetic α-tocopherol, synthetic vitamin E, or DL-α-tocopherol, and is the compound produced by condensing 2,3,5-trimethylhydroquinone with phytol, isophytol or phytyl halogenides. The reaction is carried out in acetic acid or in a neutral solvent such as benzene with an acidic catalyst, *e.g.* zinc chloride or formic acid. The product obtained is purified by vacuum distillation.

REFERENCE

1. C. Fisher and T.R. Scott, *Food Flavours: Biology and Chemistry*, Royal Society of Chemistry, Cambridge, 1997.

Chapter 5

Confectionery Plant

Most sugar confectionery is made by a process of dissolving sugar in water and boiling the sugar syrup with glucose syrup in order to concentrate the resulting mixture, and originally this was done in a saucepan on a stove. Small quantities of confectionery are usually made in this way although industrially the only products now normally made by this sort of process are those that require a temperature that is too hot for steam cooking.

In the confectionery factory the saucepan has been replaced by the steam heated pan (Figure 5.1). Steam heat provides a controllable way of heating food products – one advantage is that the maximum obtainable temperature is restricted to that of the steam. The use of steam implies a steam boiler and a system of pipes are needed to distribute the steam but the use of a central steam boiler can be avoided by using self-generating steam pans. In these devices, steam is generated *in situ* either from electric or gas heating, and using a self-generating pan avoids the capital cost of a boiler and the necessary boiler inspections and insurance.

The alternative to a central boiler is to use steam generators near to where the steam is needed. These devices, fired by gas, use a steam coil rather than a water or fire tube boiler – an advantage is that the losses in distributing steam are avoided; also, the cost of keeping a large boiler running at times of low steam consumption are avoided. Another advantage of the steam generator is that it can operate with a much higher level of dissolved solids in the feed water since dissolved solids in steam boilers tend to be deposited on the heating surfaces. This type of deposition creates problems since the boiler scale is a poor conductor of heat. This can lead not only to loss of efficiency but also to buckled boiler plates caused by thermal distortions. Dissolved solids in boilers can also cause problems with priming, *i.e.* liquid water carrying over into the steam. The level of dissolved solids in boilers is ultimately reduced by 'blowing down' the boiler, *i.e.* venting some of the boiler water. This

Figure 5.1   *Pan for making high boilings*

obviously lowers the dissolved solids; however, it is wasteful in terms of the loss of energy and water treatment chemicals.

STEAM INJECTION COOKING

One highly efficient way of producing some forms of sugar confectionery is to use a steam injection cooker. These devices inject steam under pressure into a slurry of the sugar confectionery ingredients. They were originally developed for use on starch-based products where the starch needs cooking. It has since been found that machines of this type can also be used to make gum-containing products where the machine is used to dissolve and de-aerate a slurry of sugar glucose syrup and instant gum acacia. Here, the steam and slurry can be held under pressure to cook or dissolve the ingredients in the slurry and by altering the pump rates and the back pressure the degree of cooking can be altered. When the liquids emerge from the cooker the reduction in pressure causes excess steam to flash off. The efficiency of this type of cooker comes because not only is the sensible heat used but also the latent heat of the steam. However, the disadvantage of this type of machine is that the steam has become an ingredient and must therefore be food grade, and this does restrict the boiler treatment chemicals that can be used.

VACUUM COOKING

Applying a vacuum has the effect of reducing the boiling point. The boiling point of any liquid is the point at which the vapour pressure matches that of the atmospheric pressure. Thus, lowering the atmospheric pressure allows a mixture to be concentrated at a lower temperature. The consequences of this are considerable: working at lower temperatures saves energy; it reduces heat damage; it also speeds

Figure 5.2 *Vacuum cooker*

the process up. Applying a vacuum to a mass that is currently boiling also causes a rapid reduction in water content as the water flashes into the vapour phase. The system then cools rapidly to the boiling point at the new pressure. This effect is beneficial because it saves energy (as opposed to boiling in an open pan), the cost of higher steam pressures is avoided and the product is de-aerated. Many of the mixtures that are boiled to make confectionery contain ingredients that stabilise bubbles. In a product where bubbles are not wanted, removing them can be difficult without the use of vacuum. A typical vacuum cooker is shown in Figure 5.2.

CONTINUOUS PLANT

Many process industries have converted from batch to continuous plant. Using a continuous plant is not as easy in the food industry as it is in the chemical industry since considerations of hygiene must now be added. The plant, therefore, will have to stop periodically for cleaning although continuous plants do tend to produce a more consistent product than batch processes. The most general problems with a continuous plant normally occur in setting up.

A particular confectionery industry problem is the long product life. A long-established product, initially made by a batch process, must retain the same qualities when prepared by a more modern continuous plant process, and a great deal of work can go into making the two products exactly the same. In general, however, the continuous plants are more effective at heating the product and produce less sucrose inversion, although, even after the recipe has been adjusted to account for this, textural differences sometimes persist.

Chapter 6

Sugar Glasses in the Chemistry of Boiled Sweets

Glassy materials are common in a number of areas, both natural and man-made. The humble boiled sweet is an example of a sugar glass (Figure 6.1).

The glassy state of matter is not a thermodynamic phase but a supercooled liquid, and although a glass is not a thermodynamic state of matter, glasses do exhibit a sharp transition temperature between glassy and rubbery states. Many methods have been used to determine the transition temperature between these states, and one popular method is differential scanning calorimetry (DSC). In these instruments the sample and a blank are subjected to a change of temperature, up or

Figure 6.1 *Boiled sweets (high boilings)*

down, at a controlled rate. The instrument measures the difference in energy input or energy extracted between the sample and the control and a plot of the difference shows any variation of heat capacity C_p with temperature. The glass transition is associated with a discontinuity in C_p. Unfortunately, there are a number of different ways in which the output from DSC can be analysed: some workers advocate using the beginning temperature of the discontinuity, others advocate the end temperature, whereas others advocate using the middle temperature – if DSC is being used merely to assess the difference between two samples then the absolute value matters less. Alternatively, some workers advocate integrating the result from the DSC, *i.e.* measuring the area under the curve, which then gives the variation of the enthalpy with temperature. The variation in several other physical parameters with temperature, such as the refractive index and the dielectric constant, have been used to study glass transitions. In general, however, the glass transition temperature obtained depends upon the method of measurement, and in confectionery the important point is that a product that is intended not to crystallise should be in the glassy state at ambient temperatures.

Most sugars will form a glass but pure sucrose does not. Therefore, commercial sugar glasses are always made from sucrose and some other sugar. Initially, invert sugar was used but this has now been largely replaced by glucose syrup. Any additive required to stabilise a sucrose glass is traditionally referred to as a 'doctor' by confectioners. A few recipes call for the addition of some acid to the sucrose, which generates invert sugar *in situ*, although glucose syrup is a much better stabiliser of sugar glasses than invert sugar. In practice, boiled sweets (or high boilings) become unstable by absorbing water. Initially the product becomes sticky, then soft, followed ultimately by crystallisation. The rate at which high boilings can absorb water is diffusion limited and the high molecular weight fraction in glucose syrup inhibits the migration of water into the sweet – this is why glucose syrup gives a markedly more stable product than one stabilised with invert sugar. Glucose syrup also gives the boiled mass a plastic consistency when warm which makes it much easier to handle.

THE FORMULATION OF BOILED SWEETS

The important parameter at the formulation stage is the ratio of sugar to glucose syrup. Figure 6.2 shows the ratios of sugar : glucose in common types of sugar confectionery. It can be seen that high boilings (boiled sweets) have the highest proportion of glucose syrup.

The water content of the finished product depends upon the cooking temperature, and as just mentioned in the previous section making stable boiled sweets depends upon obtaining a product with a low moisture

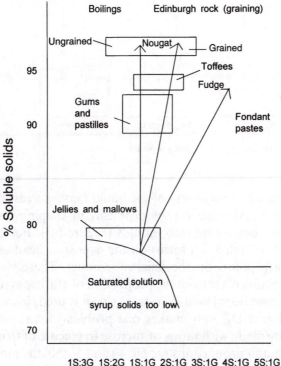

Figure 6.2 *Sugar and glucose composition of confectionery*

content. Also, the higher the glass transition temperature the more stable the product. As water can be regarded as a plasticiser it reduces the viscosity of the system and thus reduces the glass transition temperature. The important parameter at the formulation stage is the ratio of sugar to glucose syrup. Invert sugar is not normally used on its own to stabilise boiled sweets. Some formulations do contain some invert, usually as a way of incorporating rework. The situation to avoid is ending up with a product that is excessively hygroscopic, which will reduce the shelf life by making the product sticky. The problem is, therefore, to restrict the total proportion of invert sugars in the product. This includes any invert sugar that is part of the formulation, any sucrose that is inverted during cooking, and the dextrose present in the glucose syrup.

Economically, increasing the proportion of glucose syrup is beneficial; however, technically there are problems. Increasing the proportion of glucose syrup increases the proportion of dextrose and the high molecular weight dextrose oligomers. Increasing the proportion of dextrose makes the product more hygroscopic which increases the tendency of the product to become sticky. As the proportion of high molecular weight

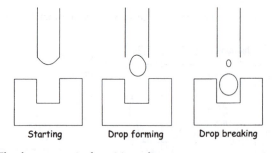

Figure 6.3   *The three stages in depositing a liquor*

material is increased the viscosity of the liquid mass increases. Excessive viscosity at the liquid stage causes problems in shaping the product, particularly when depositing into moulds (Figure 6.3). For example, if the viscosity is too high a tail forms on the depositor head and this tail can lead to sharp points on the finished product. These viscosity and hygroscopicity problems restrict the proportion of glucose syrup that can be used. If a conventional acid-converted syrup is used, incorporating a syrup of a different DE only makes one problem or the other worse. Boiled sweets are made with ratios of sucrose to glucose of from 70:30 to 45:55. The common proportions are 60:40 and 50:50, the more extreme ratios being used where special properties are required.

The acid-enzyme and particularly enzyme-enzyme syrups break away from these restrictions. In enzyme-enzyme syrups the spectrum of carbohydrates present is modified to increase the maltose while reducing the dextrose and the high molecular weight fraction. Using these syrups a high boiling can be made with a sucrose to glucose ratio of approximately 35:75 with the properties of a 55:45 product. Thus, a superior product can be made at lower cost owing to the application of enzyme technology.

MANUFACTURING PROCESSES FOR BOILED SWEETS

In making boiled sweets, the avoidance of Maillard class browning reactions is important. Although the necessary low moisture contents can be achieved by boiling under atmospheric pressure the resulting product would be an unacceptable brown colour with a caramelised flavour. The process of making boiled sweets can therefore be summarised as follows:

(1) dissolving the sugar;
(2) boiling the sugar and glucose syrup under vacuum to the final solids;
(3) cooling the boiled mass;

(4) adding flavour, colour and any acid;
(5) shaping the product;
(6) wrapping.

Small-scale Processing

On a small scale the dissolving and vacuum boiling are carried out in a steam-heated vacuum pan. An advance on this would be to use an open pan to dissolve the sugar and then to feed this to two vacuum pans. The vacuum-cooked mass is then poured onto a cooled metal slab and the mass is turned in on itself to cool the middle (Figure 6.4). Glucose syrup-based boiled sweets form a cool skin on the outside. This means that they can be manipulated by hand. Because of this skin it is necessary to fold the inside of the mass to the outside to cool it. When the mass has cooled sufficiently the colour and flavour may be added and kneaded in,

Figure 6.4   *Making boiled sweets*

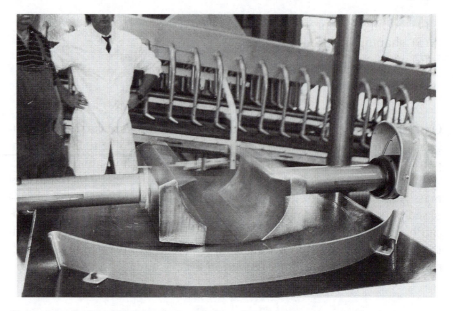

Figure 6.5   *Ruffinati boiled sweet kneading machine*

typically using machinery as in Figure 6.5. When the temperature is right the product is then fed through a drop roller. A drop roller resembles an old-fashioned mangle except that the rollers are not smooth. The rollers have depressions in them that form the shape of the finished sweets or drops. The product enters the drop rollers shapeless and emerges having been squeezed into the finished shape.

Medium-scale Processing

In medium-scale processing a dedicated batch cooker is likely to be used. There are many varieties of these cookers. The largest batch cookers work on a semi-continuous basis and are only a little smaller than continuous systems. In general, batch cookers have one chamber where the sugar is dissolved under atmospheric pressure and another where the sugar and glucose are boiled under vacuum to the final concentration. The mass emerging is then cooled on the slab as before. After cooling and flavouring the product can also be fed to drop rollers as with small-scale processes. On this larger scale the product may be fed into batch rollers which produce a rope of product. This is then fed through a system of dies that stamp the sweets into shape.

Large-scale Processing

Large-scale plants work continuously. A continuous dissolver is used to feed a continuous vacuum cooker. One improvement at the dissolving stage is that the sugar is dissolved under pressure. This increases the boiling point of the water and allows more sugar to be dissolved within a given weight of water – this reduces the energy needed to boil off the water during the second stage. The difficult part in designing a continuous vacuum cooker is in retaining the vacuum as the cooled product is removed. The cooked mass is then cooled and the colour and flavour are mixed in before the product is deposited into moulds to shape the finished sweets.

Chapter 7

Grained Sugar Products

FONDANT

The word fondant is said to come from the French verb *fondre* to dissolve. Confectionery fondants are normally made by boiling a sugar and glucose syrup in order to concentrate it before the mixture is cooled with beating to produce fine sucrose crystals. Figure 7.1 shows a typical fondant product, some peppermint creams. Some fondants used in baking are made by mixing icing sugar with a sucrose and glucose mixture. The icing sugar has a small particle size because it has been milled whereas the sucrose in a crystallised fondant has small crystals. It is possible to make a fondant using dextrose instead of sucrose.

Figure 7.1 *Peppermint creams*

Figure 7.2 *Fudge*

Crystalline dextrose has a positive heat of solution, *i.e.* heat is absorbed, and this produces a marked cooling effect which goes well with a peppermint flavour, but one which is not appreciated in other flavours. In the UK, dextrose fondants are not particularly attractive economically, but they can be in some other countries. After the fondant has been made the crystals undergo Ostwald ripening, which is the tendency of crystals to change with time. In response to small temperature variations the smallest crystals re-dissolve and the largest crystals grow.

FUDGE

A few other products involve the deliberate crystallisation of sugar. Fudge (Figure 7.2) is normally made by adding fondant to toffee or by adding icing sugar to a toffee. A very similar product can be made by making a caramel that is so formulated that it crystallises on cooling and beating. This sort of product is sometimes sold as a fudge, and in practice it is very similar.

CHOCOLATE-COVERED LIQUEUR SWEETS

While the chocolate covering of these products is not sugar confectionery, the centre is. For the centre, there are two types: the crusted liqueur that has a sugar crust (see Figure 7.3), and the crustless liqueur that has not. They are made by the following methods. The crust is made by

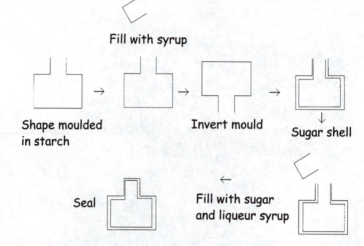

Figure 7.3 *Making liqueur chocolates*

depositing a thin layer of a sugar glucose syrup mix into a starch mould. The syrup will have been formulated to crystallise on cooling, and once cooled the layer of syrup produces a hollow mass of crystals in the mould. The problem is how to keep a liquid centre in this sugar container. This is achieved by formulating the liquid filling with a mixture of sucrose and glucose syrup that will not itself crystallise. The liquid centre is so saturated with sugars that the surrounding sugar shell cannot re-dissolve in it. Given the nature of the product it should be clear why this type of sweet is not particularly stable to temperature fluctuations.

Crustless liqueurs are made by depositing or injecting a suitable mixed sugar and alcohol syrup into a chocolate shell. The important point is that the water activity of the syrup has to be such that the centre does not dissolve sugar from the chocolate. Crustless liqueurs keep less well than crusted ones.

Chapter 8

Pan Coating

This process involves building up a coating of sugar, and sometimes of other things, layer by layer, an instantly recognisable example of a hard panned product being that of sugared almonds (Figure 8.1). Panning is one of those processes that the confectionery and pharmaceutical industries have in common.

The coating of confectionery with sugar using a pan is very ancient. Originally, these products were made in an open pan suspended over a fire by chains. Apparently the operator swung the pan by hand, coating the product with sugar syrup which was then allowed to crystallise. The normal modern small-scale method is to use a rotating dragee pan (see Figure 8.2). These are elliptical vessels made traditionally of copper, but

Figure 8.1 *Sugared almonds*

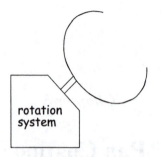

Figure 8.2   *A dragee pan*

now more commonly of stainless steel. The pan is equipped to be rotated and usually there is a system for supplying and extracting air, and possibly with a system for spraying in the sugar syrup. Typical pan sizes are between 1.5 and 0.9 m in diameter although smaller sizes are used for small-scale work.

Panned sugar coatings are divided into two types: hard panned and soft panned – a comparison of their features is given in Table 8.1. Hard panned coatings are purely sucrose, whereas soft pan coatings are a mixture of sucrose and glucose syrup. In general, hard panning is done in large pans whereas soft panning is carried out in small vessels although small products such as nonpareils tend to be made in small pans whereas, naturally, large items are more easily produced using larger pans.

The panning process can be controlled very effectively by factory employees but it is a difficult process to understand scientifically as any mathematical model involves partial differential equations.

HARD PANNING

Sugared almonds, mint imperials and nonpareils (hundreds and thousands), as well as sugar-coated chocolate lentils or eggs, are normally

Table 8.1   *Comparison of hard and soft panning*

	Hard panning	*Soft panning*
Coating chemistry	Pure sucrose	Sucrose and glucose syrup
Panning conditions	Heat and ventilation	Cold
Coating build up	Slow	Quick
Coating thickness	Thin	Thick
Pan size	Large	Small
Crystallisation caused by	Evaporation	Adding milled sugar
Typical products	Sugared almonds, mint imperials, nonpareils, sugar-coated chocolate beans	Jelly beans, dolly mixture components

hard panned although it is possible to soft pan some of these products, *e.g.* almonds. In hard panning, the centres are tumbled in the pan and a sugar syrup is applied. The rotation of the pan and the tumbling of the centres spreads the syrup out over the surface of the centres into a thin layer. The water in the sugar syrup is evaporated which causes the sugar to crystallise. Inevitably this process is slow as the water can only evaporate at certain rate and the rate of crystallisation of the sugar cannot be accelerated. The rate of evaporation can, however, be increased by increasing the temperature and the rate of air flow as well as reducing the humidity of the air. A disadvantage to this is that increasing the temperature reduces the rate of crystallisation. Once applied, the layers applied are only 10–14 μm thick, and as they are so thin they follow the contours of the product. Originally, the syrup for hard panning was added by hand as needed; modern installations use a spray system. The pans for hard panning can be heated either electrically or by steam coils although if the centre can be damaged by heat then heating the pan may not be possible. A supply and extraction system for the air flow speeds up the panning process considerably: in an unventilated pan the process slows down as the air becomes saturated with water vapour.

In order to hard pan centres they have to be coatable with a concentrated sugar syrup. Some centres, such as nonpareils, readily take a sugar syrup coating, whereas others, particularly those that have a hydrophobic surface such as nuts or chewing gum, need some pre-treatment. The stages of the hard panning process are referred as 'wetting' when the syrup is dosed in and 'engrossing' when the coating is built up.

SOFT PANNING

Soft panning involves applying a syrup to the centres in the same way as for hard panning; however, the soft panning syrup is intended not to crystallise. The syrup used here is either an all glucose syrup or a 50:50 mixture of sucrose and glucose syrup, and the centres are wetted with the syrup just sufficiently to coat them. Instead of evaporating the water as in hard panning, caster or milled sugar is then added which dissolves in the water of the syrup. The amount of sugar added needs to be just sufficient to coat the centres but must not be in excess. Any excess of sugar will be picked up in the next dose of syrup and the sugar will eventually convert the syrup from a non-crystallising syrup to a crystallising one. The centres are allowed to tumble in the pan until no more water is lost. The centres are then removed from the pan and placed on trays to dry.

Soft panning is carried out in the cold and without the use of drying air. In this case, however, dust extraction is needed for reasons of health and safety. It is a much more rapid process than hard panning and can be applied to soft centres that would be unsuitable for hard panning. Soft panning puts on thicker layers than hard panning and consequently the shape of the centre tends to be lost: three to five layers can produce a thick coat. A product that has been soft panned can be finished by dusting with milled sugar followed by a number of hard panned coats. Typical soft panned products are jelly beans and dolly mixture components.

PROBLEMS IN COATING ALMONDS AND OTHER NUTS

One of the problems in making products such as panned almonds is pre-coating the centre. The pre-coating must smooth out the irregularities in the surface, and also act as a fat barrier, particularly in the case of almonds, to prevent the escape of any nut oil. Nut oils create problems not only because they can stain the product but also because they can create adhesion and stability problems. Sugar syrup is a highly hydrophilic material and will not readily adhere to a fatty hydrophobic surface, and nut fats are the cause of problems through both oxidative and lipolytic rancidity *via* the processes outlined in Chapters 2 and 3 on pages 20–22 and 30. During processing the coating should prevent the centres from sticking together, and obviously the pre-coating must also have a neutral taste so as not to affect the taste of the product. Various materials have been used to coat the almonds: the traditional ingredient in this area is gum acacia although modified starch is also used.

GLAZING AND POLISHING

The finished pan-coated product is often treated to glaze or polish it. Glazing and polishing ingredients are either liquids or solids and are agents such as sugar, glucose syrup, starch, oils, fats, beeswax, carnauba wax, shellac, talcum, paraffin oil and paraffin wax. Natural waxes, *e.g.* carnauba, are used in the form of a fine powder and this system suits conventional dragee pans; in automatic coating machines the powder has to be distributed by a screw system. However, hard waxes like carnauba need prolonged polishing times and tend to form white spots on the surface.

Liquid polishing agents are easier to dose accurately and alcohol-based polishing agents are suitable for glazing chocolate-coated centres and for polishing sugar-panned articles. Solvent-based polishing agents give fast polishing and high gloss without spotting although the disadvantage is the need for an explosion-proof extraction system. Of course, regardless of the polishing agent used, a high gloss requires a smooth surface.

A major ingredient in glazes is shellac. This material is an insect exudate and is completely natural and biodegradable. Kosher grades are available if required.

PROCESS CONTROL SYSTEMS

In old-fashioned operation the progress of the panning operation can be controlled by an operator, one operator being capable of controlling several pans. This sort of system is very flexible and intelligent. Initially, attempts to automate panning concentrated on the time and temperature control. A more modern approach is to measure the absolute humidity in order to control the drying cycle and an example of such a system is the one produced by the Italian company Nicomatic. This is claimed to be an improvement on systems that rely solely on temperature and the elapsed time. In particular, because the actual humidity is measured, process parameters can be optimised to maximise throughput. In a time–temperature system, to avoid under-drying the time set must be the maximum necessary. This sort of system is controlled *via* a program-mable loop controller (PLC) which can be set up from a personal computer-based operator interface. The personal computer is also used for real time process monitoring. The control data cover air temperature and flow rate, drum speeds, atomising pressure and spray time as well as the humidity set point. Nicomatic machines consist of two solid coating pans with drying air brought in and extracted by two plenum blades parallel with the product flow.

AUTOMATED SYSTEMS

An alternative to automating an dragee pan system is to use a completely automated system based on a rotating drum. Typical of these machines are those made by the American company LMC International (formerly known as the Latini Machine Company). This company's APS-4 auto-matic sugar coating machine is the most recent of their equipment designs. It consists of a large, side-vented, perforated drum that rotates as the sugar is sprayed; the side perforation allows the easy passage of air through the product. The APS-4 includes air cooling, recycling, a dust collection unit and a dehumidifier. As air is passed over the centres the air evaporates moisture from them, increasing the humidity of the air. The higher the humidity of the air the less moisture it is able to remove from the centres, hence the dehumidifier speeds up the drying process and improves the quality of the finished product. The cooling unit ensures that the air is at the right temperature for effective drying, where the air used in cooling and drying is recycled to reduce the cost in providing sufficient cooled air. The sugar is dispensed through hot water-jacketed and self-cleaning stainless steel spray guns. The APS-4 is computer-controlled and only needs one operator. After programming how much sugar is to be dispensed and how often, the operator can simply walk away and leave the system to run unattended. This is radically different from the situation with dragee pans where almost continual supervision is required.

Chapter 9

Toffees and Caramels

The chemistry of toffees, caramels and fudges is the chemistry of proteins heated with reducing sugars, *i.e.* the chemistry of the Maillard reaction. Figure 9.1 shows a picture of some toffee; it has a smooth appearance compared with that of Figure 7.2, a fudge, already covered in Chapter 7. The origin of the word 'toffee' seems to be lost in time although it might be derived from 'taffy', a dialect word for 'sticky'. One authority states that toffees originally did not contain milk but were in fact high boiled products containing brown sugar, glucose syrup or invert syrup and fat, originally butter. This sort of product was a type of high moisture butterscotch and was apparently the original toffee on toffee apples – many toffee apples are now made with a coating of coloured boiled

Figure 9.1 *Toffees*

sugar. Caramel is not a defined term, at least not in the UK, however, in some other countries caramel must have a higher milk content than toffee. Confusingly, caramel is also used to refer to the brown colour that is made either by burning or chemically treating sugar. The development of Mackintosh's toffee is regarded as important as it was said to combine the attributes of soft American caramel with English hard toffee. The resulting chewy product had the advantage that it could be sold all around the world without refrigeration. Because of this history, in the UK at least, toffee and caramel seem to be used interchangeably, and toffee is used to refer to products ranging from viscous liquids to hard sugar glasses.

All toffees contain skim milk solids and usually some fat. A toffee can be made using full cream milk and butter or with skim milk and vegetable fat – some toffees lie somewhere in between the two extremes. A few toffees are extremely hard: they are in the glassy state. They are a sugar glass with some fat dispersed in it although these toffees, *e.g.* bonfire toffee, are little made nowadays. Another variation is cinder toffee which is made by heating sugar and glucose to a high temperature in the presence of bicarbonate of soda. The heat decomposes the bicarbonate of soda to give carbon dioxide; this forms a foam in the sugar glass which sets as it cools.

Most toffees are chewable rather than glassy. They are made with sugar, glucose syrup and some form of milk. The preferred form of milk for making toffees is sweetened condensed milk, either full cream or skimmed.

As toffees normally have dispersed fat in them, they are emulsions (see also Chapter 4). Toffees are nearest the oil-in-water category of emulsions since water itself is only a minor constituent of a finished toffee; the continuous phase of a toffee is a sugar and glucose syrup mix.

COOKING TOFFEES

It is perfectly possible to cook a suitable toffee using a saucepan and stove. It is said that Mackintosh's celebrated toffee started in that way in Mrs Mackintosh's kitchen. The next step in cooking technology would be to use a steam heated pan which would probably produce a slightly more uniform but just as acceptable product. If the same mixture were to be cooked in a very modern continuous plant the product would emerge off-white and with very little flavour. Colour and flavour are normally generated by localised over-heating at the edge of the pan causing Maillard reactions and if a more efficient cooking system is used this local overheating does not occur, hence the colour and flavour are not formed. Continuous toffee plants normally have special facilities to allow the toffee to develop colour and flavour.

A TYPICAL TOFFEE

A typical commercially made toffee contains the following ingredients:

Sugar

Sugar normally makes up a high proportion of the finished product: some is contained in sweetened condensed milk, and the rest goes into the recipe as crystalline sugar which is similar to the domestic granulated type. The sugar must be dissolved during processing. It would be possible to add the sugar to the mix as a 66% sugar syrup but this material is not microbiologically stable and would have to be freshly made.

Glucose Syrup 42 DE

As previously outlined on page 26, this type of glucose syrup is a major ingredient of sugar confectionery. It provides not only bulk but also a source of reducing sugar and improves the product's shelf life by lowering the water activity and resisting drying out. Economically, glucose syrup solids are less expensive than sugar.

Condensed Milk

Either skim or full cream milk is essential although milk powder can be used as a cheaper substitute.

Fat

Originally, this would be milk fat although vegetable fat is now more common. Apart from the cost advantage, a vegetable fat can be tailor-made for a particular application.

Optional Ingredients

Milk Powder

If sweetened condensed milk is not used then milk powder is the most likely alternative. In some cases, milk powder is added to sugar and water to produce a reconstituted condensed milk. One simple reason for doing this is that the plant could have been constructed to use condensed milk.

Whey Powder

This material is used to make whey toffees, acting as a substitute for some skim milk solids. It provides protein for the Maillard reaction and fat emulsification.

Hydrolysed Whey Syrup

This product is a syrup made from whey by hydrolysing the lactose to dextrose and galactose. These sugars are sufficiently soluble that the syrup can be made concentrated enough to be stable – lactose is insufficiently soluble to do this. One advantage of using hydrolysed whey syrup is that any cheesy flavours tend to be beneficial in toffee although replacing some of the condensed milk with hydrolysed whey syrup in a recipe does tend to alter the texture of the finished product.

Invert Sugar Syrup

This is not used as much as it once was because glucose syrup is cheaper and performs some of the same functions (see also page 25). It is used where the recipe was originally developed around it and cannot be substituted or where the invert syrup is recovered from some other product.

Brown Sugar

Brown sugar is another ingredient that is used less than it once was, although special grades of crystallised brown cane sugar were made specially for uses such as toffee manufacture. Some brown sugars made now are produced by adding cane molasses to the white sugar produced from sugar beet. Brown sugar adds both colour and flavour to the finished product, together with adding a small quantity of invert sugar to the product. The colour of brown sugar does not have to be declared on the label as an added colour, which is a marketing advantage.

Golden Syrup

This is partially inverted refiner's syrup from the production of cane sugar. It is sometimes used in toffees and can be regarded as being the same as adding brown invert sugar.

Emulsifiers

Emulsifiers are added to toffees to help disperse the fat (see also Chapter 4), although it is perfectly possible to make toffees that do not contain added emulsifiers if sufficient skim milk solids are present. The emulsifying effect of a considerable quantity of skim milk solids can be replaced by a very small quantity of an emulsifier, *e.g.* lecithin or distilled monoglycerides, and the price of skim milk in the EU makes this an attractive proposition. Curiously, fat that is too well dispersed can cause problems. If some fat coats the surface of the piece then the toffee will cut easily as the fat lubricates the cutting knife; if there is no surface fat then the toffee can stick to the knife. One solution to this is to have a cutting knife coated with PTFE. The PTFE has a very low energy surface that the toffee will not stick to.

Glucose Syrup 68 DE or Higher

This type of glucose syrup can be used as a direct replacement for invert sugar syrup as it has the same water activity. The attraction is that 68 DE is cheaper than invert sugar syrup unless the invert syrup is produced by recovering waste sugar. The 68 DE syrup provides a source of invert sugar as well as improving the keeping properties of the product.

Isomerised Glucose, also known as Isoglucose or High Fructose Corn Syrup

This product is a mixture of fructose and dextrose that has the same composition as invert sugar syrup. It is made by treating a high DE glucose syrup with the enzyme glucose isomerase. This converts dextrose to fructose where the degree of conversion can be varied from 0 to 100%; for example, the 50% converted syrup is a direct replacement for invert sugar syrup, the 100% converted syrup a convenient source of pure fructose. Isomerised glucose is not much used in the EU although it is a more financially attractive ingredient in the USA and Canada.

Salt

Salt is commonly added to improve the taste of a product, where the effect of salt is to round out the flavour. A use level of 0.5% is common.

Flavours and Flavourings

It is convenient to separate flavours that are added to impart a specific flavour, *e.g.* vanilla, and those that are added to, say, add the flavour of toffee. If a taste of vanilla is wanted this could be added as vanilla extract, which is natural, vanillin, which is nature-identical, or ethyl vanillin which is artificial.

Colour

The colour commonly added to toffees is caramel, *i.e.* a burnt sugar colour, although other colours are also used. Colour is added to toffees to improve their appearance. This effectively remedies the fact that very efficient toffee cookers do not brown the product as much as old-fashioned cooking methods as mentioned earlier. Instead of burning the sugar during cooking a small quantity of carbohydrate will have been caramelised under controlled conditions in order to make the caramel colour that is added to the product.

THE PROCESS

Toffees can be made using equipment ranging from an ordinary saucepan to a large continuous plant. The processes carried out are fundamentally similar.

Dissolving

Any solid sugars are first dissolved in water or a mixture of water and glucose syrup.

Emulsifying

The fat and skim milk solids are then added – the exact order in which the ingredients are added varies between manufacturers. An emulsion of the fat in the mixed sugar syrup is the end result of this stage.

Cooking

The emulsion is then cooked to achieve the final water content. If this is controlled by measuring the boiling point it should be done to a tolerance of 0.5 °C; this, for practical purposes, is 1 °F, which factory employees find easier to handle. If the cooking is done in a saucepan or steam pan the toffee can then be passed to the next stage. If the toffee has been cooked in a high technology cooker such as a wiped film or a scraped surface heat exchanger (Figure 9.2) the moisture is evaporated so rapidly that caramelisation has no time to occur. In such systems either a pressurised pre-heater or a steam-heated post-cooking holding vessel must be provided. The residence time in these vessels is then

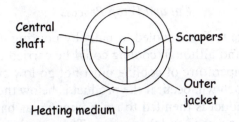

Figure 9.2 *Schematic of scraped surface heat exchanger*

adjusted to give the desired colour and flavour, and various continuous cookers have been produced with the specific aim of matching the pan-boiled product. It is possible to have a system where the pre-boil emulsion is made on a batch basis but which is cooked on a continuous basis once made.

Converting a long-established product from a batch to a continuous process can create a lot of problems. The product made on the continuous plant often has a different texture and sometimes taste to the original product: some of the differences can be traced to a lower degree of inversion in the continuously cooked product; others relate to rheological changes that occur during the cooking process.

Shaping the Toffee

The finished toffee has to be shaped in some way. Toffee is normally either run into trays, cut into slabs or used as a component of other confectionery. There are three processes used for shaping toffees for individual twist wrapping. The processes are the slab process, cut and wrap processing as well as depositing.

The Slab Process

This is a very old-fashioned way of shaping toffee. The toffee is poured, usually manually, onto water-cooled slabs. In order to facilitate manipulation of the toffee the slabs are coated with a release agent. Traditionally, this was a mineral oil although its use has been banned and long-life vegetable fats are now used. Because toffee is a poor conductor of heat it must be turned over on the slab to bring the cooler outer part into the middle and to bring the hotter middle onto the slab. If the toffee was just left on the slab to cool without turning the outside would be hard before the middle had cooled down. After sufficient cooling the toffee is cut into sheets before being cut into individual pieces.

Cut and Wrap Process

In this process the toffee is cooled, usually by pouring it onto a cooled metal drum or band although cooling could be carried out on a slab if required. The temperature of cooling must not go low enough to cause condensation, which will occur if the product is below the dew point.

The cooled product is then fed to rollers known as batch rollers that shape the mass into a cone and then a rope. The rope then passes through further rollers to reduce the thickness of the rope and then past forming wheels that produce the desired cross section. The product is cut with a rotating knife and the pieces are fed onto the wrapping paper and twist wrapped. Modern machines work at very high speeds. Although it is very quick, one problem with this type of system is that the individual pieces of toffee are distorted during wrapping.

Depositing

This is the most modern process of the three. Soft caramels can be deposited into starch moulds or chocolate shells which are later sealed by adding a chocolate back. The high technology method is to deposit the toffee into rubber moulds. The problem of making such a deposit is that the depositing machinery must be kept sufficiently hot for the toffee to flow without any further loss of water. (Such viscosity considerations are the same as seen in depositing high boilings in Figure 6.3). Toffee can thicken up if held for too long at high temperature.

In the subsequent cooling stage the problem is to abstract heat from the product and to do it without cooling it so much that condensation occurs. The rate at which heat can be removed from the finished toffee is controlled by the thermal conductivity of the toffee, which is low. After cooling, the toffees are fed to the wrapping machine as appropriate.

TOFFEE AS AN INGREDIENT OF OTHER PRODUCTS

Toffee or caramel is often used as an ingredient of other confectionery products. Examples are chocolate-covered countlines or chocolate assortments.

Formulation Considerations

If a toffee or caramel is to be used as a component it must be formulated so as make a stable product. In particular, if the toffee or caramel has a markedly different water activity from the adjacent component then problems will occur. For instance, if a bar with a cereal component, such as a wafer, is made with a caramel layer, and if the caramel does not have as low a water activity as the cereal component, then the caramel will lose water and become hard while the cereal component then gains water and softens. Any layer of chocolate will probably crack as the cereal component expands, and the consumer will probably complain and demand a refund.

Toffee in a Chocolate-coated Countline

The centres of these products are normally made as a continuous rope, either with the components previously mixed together or deposited as layers: in the case where there is a layer of caramel, the caramel passes from the cooker to a cooling drum which cools the caramel to an appropriate temperature, satisfying the requirements mentioned above; if the caramel is to be mixed with other ingredients, *e.g.* nuts, they are mixed prior to depositing. The rope of product is then cut into pieces and cooled as necessary prior to being coated with chocolate.

Toffee in a Moulded Chocolate Product

Where toffee goes into a moulded chocolate product the problems are as follows: the caramel must be fluid enough to deposit into a chocolate shell while being cool enough so as not to melt the shell; and the water activity of the caramel must be sufficiently low to keep the product stable. While these are difficult problems, they are not insurmountable.

Chapter 10

Gums, Gelled Products and Liquorice

PASTILLES, GUMS AND JELLIES

In these products there are no definitions. A jelly is normally a gel product but it can be made of any permitted gelling agent (Figure 10.1). Gums do not have to contain any gum and are commonly made of modified starch or modified starch mixed with gelatine, for example, as in Figure 10.2. Some gum sweets, however, are still made from gum acacia. When we examine the gelling agents that are used in confectionery it is apparent that chemically they are a disparate group of substances: some, *e.g.* starch, are polysaccharides whereas others, *e.g.* gelatine, are proteins.

National tastes are also different. A German gummi bear may look similar to an English jelly baby but the texture is quite different.

Figure 10.1 *Jelly products*

Figure 10.2   *Gums*

Curiously, the jelly baby was invented by an Austrian confectioner working for a British company.

Relevant Science

These products are within an area where the relevant properties are rheology, molecular structure and molecular weight. In quoting molecular weights of the biological polymers used in these materials it is worth remembering that the molecular weight distribution can be quoted on a weight or a number basis. Those properties that are colligative properties, *i.e.* boiling point elevation or osmotic pressure, depend solely upon the number of particles, and if used to determine the molecular weight will give a number average; properties such as light scattering give a weight average molecular weight. A property that is of practical importance is rheology. This affects the pumping of solutions and the depositing of sweets into moulds (see later) as well as the texture of the finished product.

Texture

Various instruments are used to study the texture of gels. Some empirical instruments are in use in the food industry but a more scientific approach is to use rheometry. A rotation rheometer is not an appropriate instrument to study gels since the rotation destroys the gel. A method

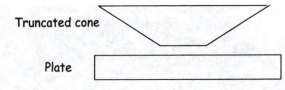

Figure 10.3   *Cone and plate geometry*

commonly used, therefore, is oscillation rheometry. In this type of measurement one element of the rheometer is oscillated while the transducer on the other element records the output. The output from this type of instrument is normally analysed in terms of two independent variables: the elasticity and the resistance. If the gel behaves as a purely elastic element, *i.e.* a perfect spring, then all of the energy input will be recovered but out-of-phase. If the gel has no elasticity but is purely resistive, all of the energy input will be dissipated. Figure 10.3 shows the cone and plate geometry used for this sort of work in instruments such as the Weissenberg Rheogoniometer. The elements are in fact a truncated cone and a plate, a geometry that gives a constant velocity gradient throughout the sample and avoids the end corrections normally needed with a concentric cylinder geometry. Originally developed as a constant sheer rate viscometer the Weissenberg Rheogoniometer can be equipped to measure normal forces, *i.e.* at right angles to the direction of rotation to be measured. It has become possible to produce constant sheer stress viscometers which can be programmed to gradually increase the stress on the sample until movement occurs – the Deer viscometer is of this type, and this sort of device can measure whether there is a true yield value. Previously, yield values were determined by extrapolating the relation between shear rate and shear stress to zero shear rate. The values obtained by this procedure were more properly pseudo-yield values.

MAKING GUMS AND JELLIES

The procedure for making gums and jellies can be generalised as in Figure 10.4. In the sections below the methods for using each type of ingredient and the differences involved are outlined.

Depositing

The product has to be deposited into moulds, either starch or starchless, where conditions must be suitable for product flow, as discussed in Chapters 6 and 9. Originally, products were hand deposited using a device like an inverted cone with a hole in the bottom. The flow of material was controlled by having a sharpened stick in the cone and

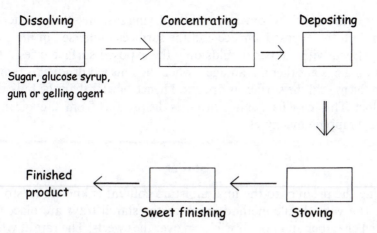

Figure 10.4 *Gum and jelly processing*

lifting or lowering the stick to control the flow. An expert with this sort of equipment could fill a tray of starch moulds fairly quickly. Factories now use machines to deposit the product into the depressions in the starch trays, and hand depositing is only done on an experimental scale in the West. Modern machines tend to do a lot of mechanical handling of the trays such as stacking of the filled trays and stamping the depressions, and all of this tends to reduce the labour content of the product.

Shaping Gums and Jellies

Traditionally these products were deposited into starch moulds, *i.e.* depressions stamped into starch, where the starch used was mixed with a small proportion of mineral oil to improve the binding of the starch. (Once again mineral oil has been replaced with a long-life vegetable oil.) The starch is dried and recycled after use.

An alternative system of starchless moulding has come into use where the product is deposited into rubber moulds. This type of system is satisfactory where the product is deposited at the final solids, *e.g.* pectin jellies. If the product has to be dried ('stoved' in the terminology of the industry – see below) to final solids then starchless moulding is unsatis- factory.

Starch moulding has the advantage that it is inherently more flexible. If the shape or weight of the product is being changed in a starchless system this requires a new set of moulds, whereas in a starch system all that needs changing is the mould board that stamps the impressions into the starch.

When a product is deposited into starch the starch dries the surface layer of the product immediately and when the starch tray is placed into the stove the starch becomes part of the drying system – the product dries from both the exposed surface and the surfaces immersed in the starch. In contrast, with rubber moulds only the exposed surface dries. Starch moulding has another advantage – when the finished product is removed from the mould the surface is drier and hence harder than the bulk of the product. This 'case hardening' protects the product from damage in high speed wrapping machines.

Stoving

Drying the product to the final moisture content is known as 'stoving'. This is a very simple method in which the starch trays are placed in a heated chamber and hot air is blown over the sweets. The rate at which a given product can be stoved is determined by the following factors: the size of the sweet; the viscosity of the sweet; and the maximum temperature that can be used without damaging the product – the rate at which the sweet can be dried is ultimately diffusion limited, *i.e.* it is limited by the rate at which water can move from the middle of the sweet to the surface.

The efficiency of stoves can be improved by dehumidifying the air and optimising its circulation. However, the air flow in the stove can only be increased to a certain point otherwise the starch would be blown around, leading to a potential explosion hazard. The actual start to finish time can be further reduced by mechanically cooling the stove, and hence the product, at the end of the cycle.

Stoves are used in a number of ways:

(1) *The air can be re-circulated.* This avoids heating cold air and only requires heat to make up for the latent heat of evaporation. The disadvantage is that the air becomes saturated with moisture, and as it picks up moisture it becomes less effective at drying.

(2) *By venting the air, either totally or partially.* This system uses more energy but maintains a lower air humidity and therefore dries more quickly.

(3) *The use of heat pump dehumidification.* This is the most effective of the systems. Here, a heat pump is used to maintain the air at a constant humidity. Changing to this system can, in some cases, reduce the drying time from three days to one day – a tray of sweets with approximately 700 g of product can be dried for one penny.

The Process for High Methoxyl Pectin Jellies

As this type of pectin sets in acid conditions at high soluble solids, pectin jellies do not need to be stoved and can be deposited at final solids.

(1) The pectin has to be dissolved in water with one third of the acid. If a high shear mixer is available then the pectin can be added directly to the water (although a small confectioner might not have such a mixer). The mixer disperses the pectin particles so far apart that they have no opportunity to self-associate. Attempting to dissolve the pectin without a high shear mixer results in the pectin forming blobs of jelly. Alternatively, the pectin can be dry mixed with five times its own weight of sucrose, where the sugar prevents the pectin gelling with itself when added to the water.

(2) The solution is boiled to dissolve the pectin completely, and the remaining sugar and glucose syrup are added.

(3) The solution is then further heated to boiling point to dissolve the remaining sugar. It is then concentrated by boiling to the desired final solids.

(4) The colours, flavours and remaining acid are added. Once this acid has been added the product can set. A slow set pectin is used otherwise the product could pre-set with disastrous consequences.

(5) The product is then deposited into moulds.

(6) The moulds are allowed to stand to allow the product to cool and set. As pectin sets under conditions of high soluble solids and low pH there is no need to stove the product, so starchless moulding can be used.

Making Gelatine Jellies

(1) The gelatine is first soaked overnight in water or possibly sugar syrup.

(2) The sugar is dissolved and mixed with glucose syrup and any invert syrup in the formula. This mixed sugar syrup is then boiled to the required soluble solids – this concentration is calculated allowing for the moisture content of the gelatine solution and any colour and flavour solution being added in the next stage.

(3) The syrup is then cooled, possibly by applying vacuum, and a solution of the gelatine, the flavour and colour is added.

(4) This material is then fed to the depositor and deposited in starch.

(5) The deposited sweets are then stoved to the required final moisture content.

The important points in using gelatine are that it is thermally labile and must not be boiled, particularly in the presence of acid, and the stoving temperature for gelatine sweets must be lower than with other gelling agents otherwise the gelatine browns.

Making Gums Using Gum Acacia

Gum acacia has traditionally been used by dissolving the cleaned tears of gum in water to make a 30% solution, which would then be allowed to stand to allow any insoluble material to settle. The gum solution would then be filtered and skimmed to remove any other insoluble matter.

Some gum merchants now supply gum in an instantised form that can be dissolved directly in cold water to make a 50% solution. The use of instant gum saves time and the energy involved in removing some of the water. However, the price of this product has to allow for the cleaning and dissolving to have been carried out, followed by a drying process.

A typical gum-making process would be:

(1) Dissolve the gum in water: this is a slow process with raw gum but can be rapid with a spray dried gum.
(2) Filter off the insolubles and skim off any floating material. As described above, instant gum would just be mixed with water as the product has already been cleaned.
(3) All of the sugar is dissolved in water and the glucose syrup is added.
(4) The mixture is now brought to the required concentration. In the simplest process the gum liquor is boiled in an open pan. As gum is a foam stabiliser the boiled liquor will have a lot of air bubbles and these can only be removed by allowing the liquor to stand for an hour before skimming off the resulting foam. This process is wasteful of time, ingredients and energy. If the liquor is boiled and a vacuum is applied the boiling point is reduced, some water boils off and the air bubbles are removed. The latent heat of evaporation also cools the liquor which reduces the cooling time. The liquor does not need to be left to stand, nor do the bubbles need to be skimmed off.
(5) The colour flavour and acid are added.
(6) The liquor is deposited into starch moulds.
(7) The moulds are stoved to the required final solids.

Alternative Methods

Steam Injection Cooking. Instant gum can be dissolved either by injecting steam directly or by making a slurry of the gum, the sugar and the glucose syrup before passing the slurry through a steam injection cooker. The steam injection cooker is not cooking the gum but merely heating the slurry and adding a little water to dissolve it. The cooker can be set up so that some water flashes off as steam when the liquor leaves the cooker and returns to atmospheric pressure. This is, of course, yet another effect produced by the applied pressure varying the boiling point.

Plate Heat Exchangers.   Alternatively, it is possible to prepare the gum liquor using a plate heat exchanger. Typically in this system, the slurry is heated sufficiently to dissolve the instant gum which has been mixed to a slurry with the sugar and glucose syrup. These plants can be configured so that the pressure of the liquor is above atmospheric to increase the boiling point and hence speed up dissolution of the gum. When the gum has dissolved the pressure can be reduced causing water to flash off as steam, thus concentrating the liquor. This system is normally used in a continuous gum-making plant. It is also possible to process gum liquors in a plate heat exchanger where the liquor is made by dissolving the raw gum in water.

In both cases the liquor must be cooled, coloured and flavoured then deposited, as in more traditional processes. While it is relatively easy to carry out all of the processes up to depositing on a continuous basis, continuous stoving is difficult – most factories operate stoving as a batch process. It is, however, possible to operate a stove in the form of a tunnel where moulds are fed in at one end and move continuously through the tunnel to emerge with the final solids content at the other end.

Making Gums Using Starch

Starch differs fundamentally from other gelling agents in that it has to be cooked in order for it to gel, *cf.* pectin and gelatine. This applies to all starches except the pre-gelatinised starches normally found in instant desserts and the high amylopectin starches. These pre-gelatinised starches are pre-cooked and hence will gel without any further treatment; the high amylopectin starches do not gel at all. The normal type of starch for making starch jellies is acid-thinned high boiling starch.

Open Pan Cooking

Starch jellies can be made by cooking the starch in an open pan, a process that requires the operator to judge when the starch is cooked. This method is now essentially obsolete as it requires a skilled worker rather than control by cooking time or temperature.

LIQUORICE

The following section covers the type of liquorice confectionery that is traditionally made in the UK. Liquorice is used to flavour other products in other countries, but British liquorice products are considered here because they are a specialised form of starch gel which is made directly from wheat flour. Inevitably, this section also covers liquorice allsorts (Figure 10.5) the traditional contents of which are:

(1) Single sandwiches, which are a layer of liquorice paste sandwiched between two layers of cream paste. Although both layers of cream paste are normally the same colour four colours are normally made – they are brown, white, yellow and pink. (Cream paste, which does not contain cream, is covered in Chapter 14.)

(2) Double sandwiches, which are two sheets of liquorice with one sheet of cream paste between them and a layer on each outer side.

(3) Liquorice plugs, which are merely pieces cut from an extruded liquorice rod.

(4) Liquorice roll, which is an extruded hollow liquorice rod with a cream paste filling.

(5) Coconut roll, which is a disk of yellow- or red-coloured cream paste which has coconut flour mixed in. In the centre of the disk is a solid extruded liquorice rod. Sometimes the coconut rolls are dipped into a drum of coconut shreds which then coat the surface.

Figure 10.5   *Liquorice Allsorts*

(6) Nonpareil jelly, which is traditionally an aniseed-flavoured gelatine jelly coated with pink- or blue-coloured nonpareils.

The flavour of liquorice is strong and persistent, and the use of liquorice root, *i.e.* the root of *Glycyrrhizia glabra*, in foods and medicines goes back hundreds of years. The plant's name comes from the Greek *glukurrhiza*, which means 'sweet root'. It is alleged that liquorice, which is a native plant to the lands around the Mediterranean, was brought to Britain by Dominican monks in the 1560s. The liquorice plant became established around Pontefract, in Yorkshire, and several factories making liquorice products were established in and around the town. The commercial growing of liquorice, however, ceased in the Pontefract area by the 1930s and liquorice root is now grown in Spain, Turkey and the Middle East.

Confectionery is not made directly from the root but from a liquorice extract. These are obtained either in the form of a spray dried powder or the paradoxically named 'block juice'. The spray dried extract is a free-flowing powder, yellowish brown in colour with a mild liquorice aroma and the characteristic bitter-sweet taste, whereas block juice is a solid block, resembling coal, but with the overpowering liquorice flavour and bitter-sweet taste.

In confectionery products liquorice is used to make Pontefract cakes (Figure 10.6), countlines, *e.g.* liquorice novelties (Figure 10.7), and pan centres, as well as the tubes and rods that go into liquorice allsorts. Pontefract cakes are a small button-like piece of liquorice; they are

Figure 10.6 *Pontefract cakes (liquorice)*

Figure 10.7   *Liquorice wheels*

sometimes called Pomfret cakes, Pomfret being an old name for the town of Pontefract.

   Liquorice is a slightly unusual example of a starch gel: instead of separating the starch, wheat flour is used directly. It is also a product where brown sugars and treacle are used. Liquorice paste is typically made from treacle, wheat flour, liquorice extract and caramel. Caramel in this context means the brown colour produced from sugar and not a form of toffee. Industrial caramel is made by the action of ammonium hydroxide on a carbohydrate, typically glucose syrup. The resulting product is not well defined chemically, and for this reason its use is recommended to be limited to 0.2% maximum.

Control of the Texture

The stages of liquorice processing are covered in Figure 10.8. Liquorice is used in a number of confectionery products which, although they are all similar, do have different textures.

   As usual in most sugar confectionery some control of the texture is achieved by varying the water content of the finished product. In liquorice it is also possible to alter the texture of the finished product by altering the degree of gelatinisation of the starch present in the flour. Traditionally, liquorice was made by cooking in flat-bottomed, well-stirred, steam-jacketed pans. These pans would have the capacity of 500 kg and it would take 2–4 hours to cook the product, the degree of

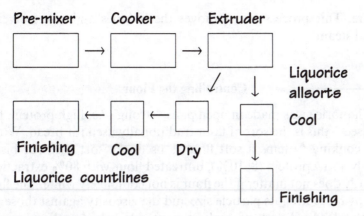

Figure 10.8 *Modern liquorice processing*

starch gelatinisation being determined by the cooking time. Such a system would use a considerable amount of skilled labour to control the process.

Nowadays it is possible to make liquorice continuously using a scraped surface heat exchanger (as seen on page 107). These machines normally consist of a heated cylinder with a central shaft with a row of floating scraper blades. In use, a premix of the treacle, wheat flour, liquorice extract, caramel, water and any rework is made. This premix is made with a moisture content some 3% higher than the desired end product. Making the premix requires care lest the flour should form lumps, and this is typically achieved by adding the ingredients according to the following:

(1) treacle;
(2) any rework;
(3) half of the flour;
(4) mix and homogenise;
(5) second half of the flour;
(6) check water content;
(7) adjust water content and recheck.

The premix must be held at a high enough temperature to be mobile but not above 65 °C as the starch would begin to gelatinise. It is then pumped to the heat exchanger where it is cooked between 120 °C and 145 °C at between 5 and 15 kg per minute. As the product is under pressure in the cooker, when it emerges to atmospheric pressure about 3% of the water flashes off. The resulting product is brown and fluffy, probably with around 18% moisture. This product is then passed to an extruder, of the sort used to shape pasta, and the paste is shaped by applying enormous pressure. This process also removes the bubbles and re-dissolves any trapped steam.

Controlling the Flour

When liquorice was made in open pans, 'strong', *i.e.* high protein, flours were used – this is the sort of flour traditionally used for bread. With the newer cooking systems a soft flour is used, the sort used for biscuits, typically a low protein (8–10%), untreated flour with 80% extraction. It obviously does not matter if the flour is not completely white. The flour is tested by measuring its particle size and the viscosity against those given in a standardised test of making a starch gel. Another important property is the amount of the enzyme amylase that is present. Cereal enzymes tend to be very resistant to the effect of heat, and the amylase in wheat is no exception. The level of amylase varies depending upon the weather around harvest time: if it rains around harvest the level of amylase will be high. If the level is too high, instead of forming a starch gel the amylase will liquefy the starch producing a sticky goo.

It might be thought that the way to measure the amylase in flour would be one of the enzyme assays used by biochemists. In practice, the Hagberg Falling Number is used. In this system, the flour is tested by mixing a weighed quantity into a slurry with water. The slurry is heated in a type of double-boiler vessel, which gelatinises the starch. The instrument then measures the time it takes a weighted plunger to descend through the starch – if there is a high amylase content the time is short. The resulting answer in seconds is the recorded parameter. In making liquorice this number should not be much less than 200.

Adjusting the Product for Different Uses

Liquorice made to final solids on a scraped surface heat exchanger (18%) will be a dull-looking, short-textured product with a matt appearance. This product would be suitable for liquorice allsorts but not for countlines. Liquorice for countlines is made by cooking liquorice to 25% water content and then drying the product to 18% moisture by placing the product in a stream of air at 65 °C. The drying time depends upon the dimensions of the product and varies between 45 minutes and eight hours. The air drying obviously allows the starch structure to build to give a less short texture. A surface gloss is produced by applying one of the following treatments:

(1) Passing ropes of liquorice through an alginate bath followed by draining and drying.

(2) Applying zein (the principal protein of maize) in an alcoholic solution of an acetylated monoglyceride followed by drying.

(3) Applying a hot solution of sugars, gum arabic and block liquorice followed by drying.

It should be noted that first and last treatments involve a polysaccharide whereas the second one uses a protein. The important property, however, is the ability to produce a good gloss rather than the chemistry of the active ingredient.

Chapter 11

Chewing Gum

Chewing gum is traditionally a mixture of gum base, glucose syrup and sucrose, and is the product where confectionery becomes involved with polymer chemistry. The gum base in chewing gum is technically a rubber, and the original ingredients were the natural products chicle and jelutong-pontianak. These are used either alone or in a mixed gum base. Some other natural substances that have been used are gutter siak, lache capsi, perillo, nispero tinu and pine tree resin. The synthetic polymers used are mainly vinyl esters, in particular poly(vinyl acetate). Styrene–butadiene rubber, isobutylene–isoprene copolymer and polyisobutylene can also be used. As in other fields of polymer chemistry, plasticisers are used, such as glycerine esters of hydrogenated or polymerised resin, lanolin, and potassium and sodium stearates. Nowadays, most manu-facturers of chewing gum buy in their gum base from a specialist manufacturer.

Different gum bases are used for chewing gum and bubble gum. Bubble gum base contains either higher levels of polymers or polymers with a higher molecular weight. Both of these foundations make the gum base more extensile and hence able to form bubbles.

Special non-sticking gum bases have been developed in order to avoid the problem of discarded gum becoming a nuisance. These products are very different to ordinary gum bases.

GUM BASE CHARACTERISTICS

The characteristics of the chosen gum base that is used depends not only upon the chewing properties of the finished product but also upon the type of gum being made. For example, chewing gum that is being pan coated has to be more rigid than gum that is being presented as a stick. If too soft a gum base is used on a pan-coated gum, instead of the product being presented as a neat pillow shape it will be distorted.

TEXTURISERS

Texturisers are substances that are added to the gum base to modify the mouth feel and facilitate processing. Common texturisers are calcium carbonate or talc – both substances obviously have to be of food grade. As both talc and calcium carbonate are less expensive than the other ingredients in gum base, not surprisingly, this leads to low cost chewing gum bases containing 45–55% texturisers whereas for a high quality chewing gum base the texturiser levels are 18–20%. Similarly, bubble gum bases have a texturiser level varying between 30 and 60%, the higher levels being found in the most economical grades. As can probably be expected, gum bases with the higher levels of texturiser place more constraints on the rest of the formulation if a satisfactory product is to be made.

Calcium carbonate is not an acceptable texturiser in products where there is an acid component to the flavour – obviously, the calcium carbonate reacts with the acid to produce carbon dioxide. Typically, acids are only used in fruit-flavoured products, and here, talc must be used as a texturiser. If, by accident, calcium carbonate is used the carbon dioxide generated could blow up a sealed package causing the packaging to fail. The acid flavour is also lost through the acid–carbonate reaction.

ANTIOXIDANTS

Gum bases normally contain a permitted antioxidant. Typically, butylated hydroxytoluene (BHT), butylated hydroxyanisole (BHA) or tocopherols are used. The reason that these substances are used is that the gum base is subject to oxidation, and antioxidants work because they act as free-radical traps. Oxidation normally occurs *via* a free-radical mechanism, and because of this, oxidation is a zero free energy process and relatively unaffected by the ambient temperature.

SUGARS

If the finished product is not to have a gritty feel in the mouth the sugar has to have very fine particle size. Individuals differ in their response to particle size but particles above 20–40 μm are normally judged to be gritty. Chewing gum has, therefore, to be made from milled sugar, a product much like icing sugar. For instance, a typical specification for sugar for use in chewing gum would be 4% retained on a 200 mesh screen, *i.e.* 96% passing through the sieve. A major chewing gum manufacturer will take in crystalline sugar and mill it on-site and feed the milled sugar directly into the chewing gum production. Milled sugars are difficult to handle since they are an explosion hazard; also if exposed to high humidity they agglomerate. If introducing the sugar directly into the gum during production is not possible then 3% of dried starch can be introduced, ideally at the milling stage.

Dextrose

Dextrose monohydrate is sometimes used as an alternative to sucrose in chewing gum, and in some countries this substitution is economically advantageous. The endothermic heat of solution of dextrose gives a cooling sensation in the mouth, a property that goes well with mint flavours but not with others.

Glucose Syrup

The glucose syrup used in chewing gum is normally confectioner's glucose of about 38 DE, the only special requirement being that the sulfur dioxide level of the syrup should be less than 40 ppm. Glucose syrups intended for use in confectionery can contain up to 300–400 ppm SO_2; although in most other sugar confectionery products, *e.g.* boiled sweets, the glucose syrup is boiled, during which any SO_2 is boiled off. Obviously, these syrups are not suitable for use in chewing gum. The 40 ppm limit for the syrup arises because a chewing gum can contain 25% glucose syrup and the finished product should not contain more than 10 ppm of SO_2.

LOSS OR GAIN OF MOISTURE

Chewing gum can be spoiled by either loss or gain of moisture: if the gum picks up too much moisture it will become too soft and could darken; if the gum dries out it becomes too hard. Either of these problems can be prevented by wrapping the product in a moisture-proof barrier although, in practice, wrappings are not always moisture-proof and some products are traditionally presented in twist wraps which are not moisture barriers. It is common to add humectants to chewing gum in order to lower the water activity and hence reduce drying out; common humectants in this application are glycerol and sorbitol. Whereas sorbitol is a purely vegetable product, glycerol can be produced by the hydrolysis of fat. This includes animal fats, which can cause problems with some religious and ethnic groups. Glycerol can also be produced as a product from petrochemical origins.

FLAVOURS

The flavour is obviously an important component of chewing gum, and common flavours are normally compounded from essential oils. Essential oils give good flavours but they also affect the texture of the chewing gum since they act as plasticisers. The other component of fruit flavours is fruit acids. This of course precludes using calcium carbonate as a texturiser.

Chapter 12

Aerated Products

Aerated products can be viewed as selling air for money. This is not entirely true as to produce an aerated product normally requires some expensive ingredients. Technically, the problems are in making and stabilising a foam (see also Chapter 4), and typically, aerated confectionery products involve making a foam and then causing it to set. To achieve a suitable foam a popular ingredient is egg albumen, which is a whipping agent and can be set irreversibly by heat, although some products are made with a combination of egg albumen and a gelling agent, *e.g.* gelatine. The only aerated confectionery products that work differently are those that are high boilings that are chemically aerated and rely upon the product passing through the glass transition before the bubbles collapse. These products sometimes contain material such as gelatine to enhance foaming. Examples are bonfire toffee, honeycomb crunch and peanut brittle.

MARSHMALLOWS

These products were originally flavoured with the herb marshmallow but a modern marshmallow (Figure 12.1) is a foam that has been stabilised by a gelling agent. The traditional material for marshmallows is egg albumen which acts both as a foaming and gelling agent – this gives a light, soft gel. Gelatine is now more commonly used, and once again can act as both a whipping and gelling agent. A gelatine marshmallow is soft and rubbery but heavier than an egg albumen one. Further gelling agents, *e.g.* pectin, agar and starch, all give a soft, short textured gel. These gelling agents have the advantage of being vegetable products but have the disadvantage of normally needing a whipping agent as well. Enzyme-modified soya products have also recently become available for this use.

Figure 12.1   *Marshmallows*

Possible Methods for the Manufacture of Marshmallows

Batch Method

The sugar, glucose syrup and any invert sugar are boiled to temperature, cooled, and a solution of the gelling agent of choice is added. The mixture is whipped as required and then deposited in starch moulds.

Continuous Manufacture

In this type of process a mixture of sugar, glucose syrup and possibly invert syrup is boiled before being cooled to 66 °C. (A scraped surface cooler might be used here.) The gelling and whipping agent or agents are then added and the mixture is fed to a continuous whipping machine where the aerated product is being coloured and flavoured, air or gas at high pressure being used for aeration. The mix is then extruded to atmospheric pressure which causes it to expand. The rope of product may then be laid on a conveyor belt, cooled and cut to length. This produces ropes of mallow, Figure 12.2, although other shaping technologies may be used.

NOUGAT

There are a range of products sold as nougat. The original product was *Nougat de Montelimar* which was made with eggs, sugar and honey, and

Figure 12.2   *Mallow cables (aerated product)*

had almonds, cherries and angelica added. (Some confectioners would argue that montelimar is a separate product.) The dark montelimars that are common are made by adding cocoa powder, usually to a less aerated confection.

Nougat can be made either in batches or continuously although the best nougat is made in batches. Various whipping agents can be used such as egg albumen, gelatine, milk protein and enzyme-modified soya protein. Starch or gum arabic can also be used in addition. The composition can be adjusted to give the desired texture.

In continuous processing the whipping agents are beaten into a hot, mixed sugar syrup. The product is then extruded, cooled and cut to shape.

In the batch process a mixture of sugar and glucose syrup is boiled to 8% moisture content, typically using a vacuum pan. The syrup is then transferred to a whipping machine and the whipping or gelling agent or agents are added – the traditional material is egg albumen. The product is whipped with the egg albumen which is set by the heat, and the fat, nuts, fruit and possibly a small amount of milled sugar are then added. The product is poured into trays lined with rice paper and are left overnight to allow some of the sugar to crystallise. The product can then be cut into shape and wrapped.

Chapter 13

Sugar-free Confectionery

More science has gone into making products that resemble sugar confectionery but which are not based on sucrose and glucose syrup than ever went into developing traditional confectionery products. Although this is obviously an oxymoron, there is a tendency to call such products sugar-free sugar confectionery; and a better name might be Sugar Confectionery Analogues. This chapter covers products formulated to satisfy special dietary needs, *i.e.* those made to satisfy health food and reduced calorie claims, and sugar-free or special products that are suitable for diabetics (Figure 13.1). These are a diverse set of requirements.

Special products intended for diabetics have been made for many years, the aim of these products being that they should not cause a surge

Figure 13.1 *One calorie controlled and two sugar-free products*

131

of blood sugar. With this aim in mind, diabetic products normally exclude sucrose and glucose syrup. Fructose, although it is chemically a sugar, is metabolised independently of insulin and is generally accepted as suitable for diabetics. The commonest ingredients in diabetic confectionery are the polyols. These substances are only slowly absorbed, thus avoiding any surge in blood glucose levels. There are also a few individuals who, because of a metabolic problem, are unable to consume sucrose.

Another reason for consuming sugar-free products is the belief that refined sugar is in some way unhealthy. 'Tooth friendly' claims are highly specific: the product must be tested to see that it does not cause a fall in pH during eating, and this system requires the use of a specialised pH electrode strapped to the teeth of a volunteer. To pass this test the product must be free of any fermentable carbohydrate and acids – this does have the odd effect in that some products which contain concentrated fruit juice fail the test and have to be re-formulated. The calorific values accepted by the authorities are not universal. The current position in the European Union is that the polyols are only partially absorbed to the extent of 2.4 kcal g^{-1} as opposed to 4 kcal g^{-1} if they are completely absorbed. For example, the polyol lactitol has the following accepted values:

EU, 2.4 kcal g^{-1};
USA, 2.0 kcal g^{-1};
Canada, 2.6 kcal g^{-1}.
Japan 1.5 kcal g^{-1} for lactitol monohydrate, and 1.6 kcal g^{-1} for anhydrous lactitol.

Crossing the border between the USA and Canada must have a considerable effect!

LAXATIVE EFFECTS

One problem with all of the polyols except erythritol, and to a lesser extent with polydextrose, is that they can have a laxative effect. This effect is osmotic in origin where the unabsorbed material upsets the osmotic balance within the gut; the necessary correction can have unpleasant consequences. Although values for acceptable consumption are published, the response of individuals varies considerably; a few individuals are hypersensitive to this effect, whereas some other people are largely insensitive to it. Warnings to this effect are required in the UK, and the laxative threshold has to be the most effective limit to consumption yet devised. Erythritol avoids the laxative effect because it is excreted *via* the kidneys.

THE SUGAR SUBSTITUTES

Bulk Sweeteners – The Polyols

The ingredients usually used in these products are polyols such as sorbitol, maltitol, lactitol, isomalt, erythritol and polydextrose:

Sorbitol (**1**) is hydrogenated glucose.
Maltitol (**2**) is hydrogenated maltose.
Lactitol (**3**) is hydrogenated lactose.
Hydrogenated Glucose Syrup is made by hydrogenating a high maltose glucose syrup.

These products are a special enzyme-converted glucose syrup which is then hydrogenated.

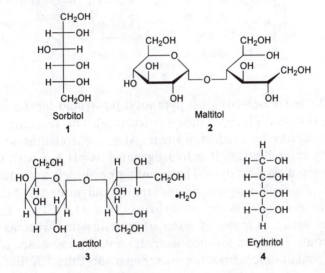

Maltitol

This material (**2**) is produced by hydrogenating maltose in the form of a pure maltose glucose syrup. Maltose has the advantage that like sucrose it is a disaccharide and it is one of the sugar substitutes that is closest to sucrose. Legally, it is a special case of a high maltitol hydrogenated glucose syrup.

Erythritol

Although erythritol (**4**) is a polyol it avoids the laxative threshold problem because it is excreted *via* the kidneys. Unfortunately, erythritol is not legal in the EU at the time of writing (1999) although it is legal in the USA and some other countries. There appears to be no good reason why erythritol should not gain approval in the EU, and at the rate that approvals are granted it will probably be the 21st century before this happens. The calorific value for erythritol is only 0.2 kcal g^{-1}.

Isomalt

It is not possible to make a polyol directly from sucrose since the reducing groups are not accessible. However, by some clever chemistry isomalt (**5**) is produced from sucrose. Sucrose is first rearranged

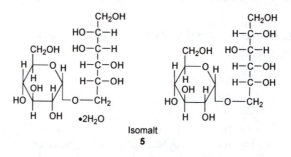

Isomalt
5

enzymically from 2-*O*-α-D-glucopyranosyl-*β*-D-fructo furanoside to iso-maltulose (6-*O*-α-D-glucopyranosyl-D-fructose). This rearrangement has converted sucrose to a reducing sugar. After the isomaltulose has been purified by crystallisation it is hydrogenated at pH 6–8 with hydrogen and a Raney nickel catalyst. The resulting product is a mixture of the isomers 1-*O*-α-D-glucopyranosyl-D-mannitol dihydrate (1,1-GPM dihy-drate) and 6-*O*-α-D-glucopyranosyl-D-sorbitol (1,6-GPS). The GPM crystallises with two moles of water of crystallisation whereas the GPS is anhydrous. Thus the finished isomalt contains 5% water of crystal-lisation. Isomalt differs from the other sugar substitutes in that it is non-hygroscopic. Chemically, it is also extremely stable.

Polydextrose

Polydextrose (**6**) is a synthetic polymer of dextrose with the dextrose molecules linked 1→6. This link is not common in nature so polydex-trose is only one quarter metabolised which gives the psychologically important value of 1 kcal g^{-1}. As laxative effects arise through colligative properties, and because polydextrose has a high molecular weight, the laxative effects experienced are lower than for the low molecular weight polyols.

When polydextrose first became available the material supplied had a considerable quantity of citric acid as an impurity as well as some bitter impurities with a brown colour. The forms commercially available now

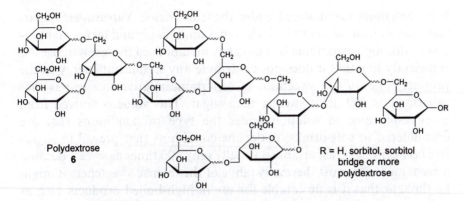

Polydextrose
6

R = H, sorbitol, sorbitol
bridge or more
polydextrose

are considerably more pure and are free from the bitter compounds and citric acid. These products are available under the trade name *Litesse II*. A further variety of the product is *Litesse III* which is normally available as a syrup. In this version the reducing groups have been hydrogenated to produce a polyol polymer.

Intense Sweeteners

Of the bulk sweeteners above, only maltitol or hydrogenated glucose syrup produce products that are sweet enough to be near to conventional sugar confectionery. The difference in sweetness has to be made up by using an intense sweetener, the substances below are used in this application. As some of these substances are not universally legal the use of intense sweeteners is strictly regulated and in general, intense sweeteners are regulated on a permitted list basis, *i.e.* the substance is illegal unless it is specifically permitted. For example, cyclamates have been banned in English speaking countries but have remained legal elsewhere, and stevioside is legal in Japan but nowhere else.

Aspartame

Aspartame (7) is a dipeptide sweetener. Chemically it is *N*-L-α-aspartyl-L-phenylalanine 1-methyl ester, with a molecular formula of $C_{14}H_{18}N_2O_5$.

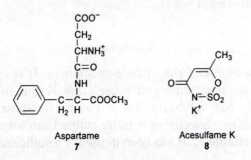

Aspartame
7

Acesulfame K
8

It is sometimes encountered under the trade name *Nutrasweet*. Aspartame has a clean sweet taste which comes out quickly and leaves no aftertaste. Although aspartame is completely metabolised the levels of use are sufficiently low that it does not contribute any significant energy to the product. The quoted sweetness of aspartame relative to sugar is that aspartame is 200 times sweeter than sugar. This value is derived from using aspartame in solution under the type of conditions that are encountered in soft drinks; under the conditions that prevail in sugar-free confectionery, aspartame is usually only 100 times as sweet. Because aspartame is the most thermally labile of the intense sweeteners it might be thought that it is unsuitable for use in high-boiled products such as boiled sweets. In fact, aspartame is very successful in this application. Boiled sweet processing of conventional products is adapted so that the flavour, colour and acid can be added while the product mass is cooling but before it becomes too stiff to handle. Adding the aspartame at a similar stage is very successful; a little aspartame is lost initially but afterwards the aspartame remains unchanged. The important point is that the very low water activity and very high viscosity of the sweet mass combine to preserve the sweetener. In other branches of chemistry, labile substances are sometimes trapped in a glass to permit further study. In this case, the glass is a sugar-free boiled sweet and it forms at a high temperature.

Acesulfame K

This intense sweetener (**8**) is quoted as having the same effective sweetness as aspartame, but unlike aspartame it is sufficiently heat stable that it can be added at the beginning of the boil in high-boiled products. If a product with the same amount of acesulfame K is compared with one based on aspartame the taste will be different. In practice acesulfame K is not normally used on its own but is sometimes used with aspartame. Chemically, acesulfame K is the potassium salt of 6-methyl-1,2,3-oxathiazine-4(3H)-one-2,2 dioxide or 3,4-dihydro-6-methyl-1,2,3-oxathiazine-4-one 2,2-dioxide. It can be regarded as a derivative of acetoacetic acid. The empirical formula is $C_4H_4NO_4KS$ and its molecular weight is 201.2.

Saccharin

Saccharin is the oldest of the intense sweeteners. It is again quoted as being as sweet as aspartame and acesulfame K but in practice most individuals find that saccharin has a bitter after-taste which is not regarded as pleasant. Saccharin is more stable than aspartame but less stable than acesulfame K. It has been in use for a sufficient length of time that it was possible to prove that there was no increased cancer incidence in high saccharin users than the general public. Saccharin is not metabolised at all.

Stevioside

This sweetener is extracted from the Peruvian plant *Stevia rebaudiana*. It is legal in Japan but not elsewhere.

Thaumatin

This protein sweetener is extracted from the berry of a fruit which grows in west Africa. It has been sold under the trade name *Talin*. Thaumatin is intensely sweet. The sweetness is slow in onset but is long lasting with a liquorice end-note. It is little used in confectionery although it has been used as a coating on the outside of chewing gum where its persistence would be an advantage.

Neohesperidine Dihydrochalcone (NHDC)

This intense sweetener is made from grapefruit skins and it has a liquorice-like sweet taste. NHDC (**9**) has long been suggested as a potential intense sweetener but has only recently received legislative approvals. It is 900 times sweeter than sucrose, and chemically NHDC is the open chain analogue of neohesperedin a flavonone which occurs in Seville (bitter) oranges (*Citrus aurantum*). The dihydrochalcones are flavonoids which are ubiquitous in plants; flavonones, chalcones and anthocyanins are also flavonoids.

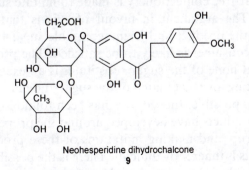

Neohesperidine dihydrochalcone
9

   Like thaumatin, the taste of NHDC has a slow onset, and an after-taste of menthol or liquorice.

Sucralose

This intense sweetener is a chlorinated sucrose. It is legal in Canada but not elsewhere at the time of writing. Although chemically derived from sucrose it is not metabolised.

Synergy

Where intense sweeteners are used, claims are made of synergy between sweeteners, *i.e.* a blend of sweeteners provides more sweetness than the sum of the amount provided by each singly. There is evidence of this effect in soft drinks although in confectionery evidence is less apparent. Synergy is not just a crude numerical effect. There can be synergy between a rapid onset sweetness as in aspartame combining with another sweetener with a slow onset sweetness.

The Chemistry of Sweetness

Chemically, these intense sweeteners are a very diverse group of substances. The early forms available were discovered serendipitously, usually by an experimenter accidentally tasting some. All of the intense sweeteners appear to be able to fit onto a particular receptor on the human tongue, and quite small chemical changes in their structures can convert a substance from sweet to bitter or *vice versa*.

It should be pointed out that the response of individuals to intense sweeteners is variable whereas the response to ordinary sugar is not.

MAKING SUGAR-FREE PRODUCTS

In general, sugar-free confectionery is made to imitate sugar-containing confectionery. The argument in favour of this is that the consumer already knows the product and if an acceptable sugar-free analogue is made then the consumer is likely to be satisfied. The problem with this approach is that none of the sugar substitutes is an exact substitute for sucrose, so making an exact match of the sugar and glucose product will probably not be possible. Indeed, this has been found in practice: sugar confectionery products have developed around the properties of sucrose and glucose syrup, and making imitations of these products but using other ingredients is inherently difficult. There is the possibility of making products that are based around the inherent properties of the sugar-free ingredients, which indeed could not be made with sucrose. There is, however, little evidence of this happening.

REDUCING THE ENERGY CONTENT

A popular claim, particularly in the English speaking countries is the term 'low calorie'. Calorie reduction in most products is made by one of a number of ways: replacing ingredients with water; lowering the fat content; reducing the density; or making the portion smaller. In sugar confectionery, replacing ingredients with water is not really possible without making the product unstable. Some sugar confectionery products do not have any fat in them anyway, whereas in others, because fat is more expensive than sugar, the fat content is already near the technically necessary minimum. The density of the product can be reduced by aerating it.

A further way of making energy reductions lies in replacing the sugar and glucose by either polyols or polydextrose – on paper, a total replacement of carbohydrates with polydextrose should produce a product with only 25% of the energy content of the original product. In practice, however, polydextrose is so dissimilar, particularly with regard to molecular weight, that the resulting product bears little relation to any confectionery based on sugar and glucose syrup. For example, although polydextrose forms a glass, and hence can be made into boiled sweets, its high molecular weight means that a stable finished product is made with a moisture content of 5%, compared with the fractions of 1% normally found. Practical products, therefore, have to be based on a mixture of sugar replacers. In sugar confectionery, a mixture of sugars is normally used because the solubility properties of sucrose force the use of more than one sugar. In sugar-free confectionery, many of the older ingredients, particularly the polyols, are sufficiently soluble that a stable product is possible using only one sugar substitute. However, it has emerged empirically that some of the most successful sugar-free products are made using a mixture of sugar substitutes, although the promotion of one-substance products has been encouraged because often the different ingredients have different suppliers and the suppliers have been unwilling to co-operate.

SUGAR-FREE PRODUCTS

Chewing Gum

Sugar-free chewing gum has been the biggest success of any sugar-free product, and the tonnage of sugar-free gum sold now exceeds that of sugar-containing gum. The laxative threshold problem does not affect chewing gum very much, probably because chewing, by definition, means that the product cannot be consumed quickly.

Boiled Sweets

If a sugar-free high boiling is to be as sweet as a sugar- and glucose-based product then intense sweeteners must be added. The choice of intense sweetener has to be made on the basis of a number of technical, legislative and financial considerations.

As mentioned on pages 135 and 136, aspartame might be thought to be too heat sensitive for this application, but in practice this is not so if aspartame is added to the process as late as possible. All other heat sensitive ingredients like flavour, colour and acid have to be added while the mass is still sufficiently warm to be flowable but not too hot to cause decomposition. In results reported in *Kennedys Confection*, less than 5% of the aspartame was lost upon addition; losses of aspartame during storage were found to be equivalent to only a 2% loss of sweetness in a year. Presumably this is because of the low water activity in the finished product. The clean, quickly released sweetness of aspartame works well in these products.

Acesulfame K is, in contrast, sufficiently heat stable to be added at the beginning of the boil. Unfortunately, the sweetness profile is not as good. One approach to using intense sweeteners is to use a combination of ingredients, and common combinations such as aspartame and acesulfame K or aspartame and saccharin are used. Where they are legal, cyclamates are also used in this application.

Sugar-free boiled sweets have been relatively successful. Although many polyols will form a glass, the substance that has been very successful is isomalt and around 50% of the sugar-free high boilings on the market are based on isomalt. This is the area where isomalt high boilings perform the best. They have the advantage that they are very stable, not only compared with other sugar-free high boilings but also compared to the standard sugar–glucose product. For example, an isomalt high boiling held at 63% relative humidity (RH) and 20 °C increases in weight by less than 2% after 70 days; even at 76% (RH) and 20 °C the increase in weight is still around 2% after 70 days. These conditions would rapidly destroy sugar and glucose syrup products. Hence, provided that the water content is below 2%, very stable sweets are produced.

The Problems of Making Sugar-free High Boilings from Isomalt

The inherent technical problems of making high boilings from isomalt are considerable but they have been solved. Isomalt has only 45–60% of the sweetness of sugar, and therefore the reduced sweetness is normally made up by adding an intense sweetener. Also, the solubility of isomalt at 20 °C is only 24.5 g per 100 g water; thus at low temperatures isomalt and water are handled as a slurry. In the temperature range between 90 and 180 °C isomalt masses are less viscous than a sugar and glucose syrup mass and this lower viscosity tends to cause problems in plants designed to process sucrose and glucose syrup high boilings. A further related problem is the need for more cooling, needed because the mass could be hotter and because the heat capacity is approximately 17% higher than for a sucrose and glucose syrup high boiling. The low viscosity of the mass can cause problems with the discharge system from the vacuum chamber of some plants.

Not too surprisingly, a high boilings process plant, developed to make sucrose and glucose syrup products, is not optimised to produce sugar-free high boilings. The makers of isomalt therefore recommend the following modifications to a typical system. First, the throughput should be reduced by 50% to ensure the longest possible residence time in the vacuum chamber. Then the draw off rollers used to pull the product out should only be heated with steam at 0.5 bar rather than the 3 bar steam pressure normally used. Also, the vacuum chamber should only be filled one third full to obtain optimum turbulence. These recommendations show the considerable difficulties in producing isomalt high boilings.

Typically, as an example, an isomalt high boiling needs a final moisture content of 1–1.5% for stability of the finished product. To achieve this moisture content, standard machinery requires 165 °C (330 °F) followed by an extended vacuum process. Additional cooling from the higher temperature is also needed and unless this cooling is provided a reduced throughput will be obtained, which can be as much as 50%. In some factories, steam pressure is limited to 85 psi (6 bar), which is not high enough, and installing a higher pressure steam plant is a considerable capital cost.

The low viscosity of the boiled and vacuum-treated mass also tends to cause hold-ups when the mass is removed from the vacuum chamber. Sweet-shaping machinery is designed to handle conventional boiled sweets and cannot work with a much less viscous mass. A system which uses boiling under reduced pressure at 140 °C (285 °F) can now be installed. This is the same temperature as used in the same firm's standard plant for sugar and glucose syrup products. Because this system is passing all the water vapour that is boiled off through the vacuum system, a more powerful vacuum pump is needed than that fitted to the standard system. This is the only capital expenditure required. Because the mass is cooler in the vacuum chamber, less cooling is needed. This process is claimed to give an isomalt mass which is nearly as viscous as a sucrose and glucose syrup one. Conveniently for manufacturers it is possible to retrofit existing plants.

Gums and Jellies

Sugar-free gums and jellies are relatively easy to make although some work needs to be done to adjust the manufacturing conditions because the molecular weight of the substitute differs from that of the mixture of sugar and glucose normally used. Although pectin is a polysaccharide it is normally stabilised by adding sugar: this must be considered if pectin is to be used as a gelling agent in these products. If gum acacia is to be used it should be remembered that most jurisdictions classify it as being 50% metabolised. The accepted calorific valued for this substance has varied from 0 to 100% over the years.

Turkish Delight

It is possible by careful choice of ingredients to make a Turkish delight with a reduced energy content yet resembling the full sugar version that is commonly sold in the UK.

Toffee

One problem arises with sugar-free toffees. What about the lactose? Lactose is a native component of skim milk solids and if the product is to be sugar-free then lactose-free milk powder will have to be used. If lactose is acceptable, as it would be in a reduced calorie product, then by replacing the sugar and glucose acceptable products can be made.

CONTROLLED CALORIE PRODUCTS

A number of products have appeared on the market that consist of a chocolate-covered bar with a centre composed of expanded cereals with a caramel or toffee centre (Figure 13.2). These products make the claim that the calorific value of the entire product is in the region of 90 kcal per 100 g. This of course is just below the psychologically important value of 100. The sugar confectionery products in the centre normally use polydextrose in order to reduce the calorific value. Montelimar centres are whipped which, although it does not reduce the calorific value per gram, does reduce the density. The reduced density then reduces the weight of the product which automatically reduces the calorific value. As the claim is made on a per serving basis this is perfectly legitimate. This idea has been taken further in the United States of America where M&M Mars Inc has produced a reduced calorie version of their *Milky Way* bar (this is similar to the product

Figure 13.2 *Centre of calorie controlled bar*

sold in Europe as the *Mars* bar, not the product sold in Europe as *Milky Way*). This reduced calorie product is smaller than the standard product and contains polydextrose. Both the standard and the reduced energy product retail for the same price.

Chapter 14

Lozenges

Lozenges are one of the oldest forms of sugar confectionery, and as they do not involve boiling sugar they are one of the few forms of sugar confectionery that can readily be made at home. Lozenges are simply cut from a sheet of dough and are then dried.

Lozenge making is another area where sugar confectionery is related to a different area of chemistry, this time to pharmaceutical products. Lozenges (Figure 14.1) should not, however, be confused with tablets, which may appear similar but are in fact made by a totally different process.

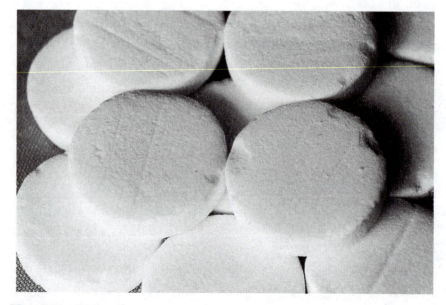

Figure 14.1 *Lozenges*

MAKING LOZENGES

The general process for making lozenges is shown in Figure 14.2. Lozenges are made by taking a milled sugar, like icing sugar, and making it into a dough – the best textures are achieved by using sugar of the smallest possible particle size. Commercial lozenges are normally made with a binder, the traditional binder being gum tragacanth which is probably the best material. Other materials that are used are gum acacia, gelatine and xanthan gum although a mixture of gum acacia and gum tragacanth is sometimes used. Gum arabic is not very suitable as used alone it gives very brittle sweets.

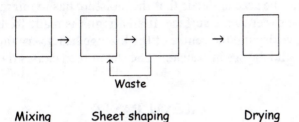

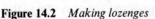

Mixing   Sheet shaping   Drying

Figure 14.2   *Making lozenges*

Gum tragacanth needs to be soaked in water to dissolve and build up its viscosity and, ideally, 12 hours should be allowed although most of the viscosity build-up occurs within the first two hours. Gelatine needs to be soaked before making it into a solution and both gum acacia and gum tragacanth are commonly used in an instantised, possibly spray dried, form. A common procedure is to have a mixture of gum acacia and gum tragacanth that have been spray dried together, and using this material gives the factory operatives only one material to weigh and make up. Regardless of which agent is used, binder solutions need to be handled with care as they could become contaminated microbiologically.

The lozenge dough is normally mixed using a powerful, slow 'z' blade mixer. The mixed dough should be firm and doughy, and when squeezed by hand a correctly made dough should stick together but not exude water. Excessive addition of water will give a sticky dough that will crack upon drying, and the sugar will also start to dissolve rather than staying as a dispersion in the binder. The dough is next rolled into a sheet and pieces are stamped out from it. In a factory the dough is likely to be passed through rollers in order to produce the correct thickness.

DRYING

The stamped pieces are spread out onto trays where a slight crust of dry sugar forms on the surface. The lozenges are then dried at 35–40 °C until the moisture content is reduced to around 1.5%. Lozenge drying is difficult: if the lozenges are dried too rapidly then they shrink and crack, and extended drying times lead to loss of flavour and deterioration of the colour. A trace of blue colour is often added to notionally white lozenges – this produces the same effect as with the optical brighteners used in washing powders. From a consumer's perspective, the effect of a blue-grey colour is generally more pleasing than a yellow-grey. Flavours are a problem in lozenges since they must be volatile enough to be perceived in the finished product yet not too volatile so as to be lost during the drying process. (In practice, lozenge makers have to accept that some of the flavour will be lost during drying.) The reason why drying cannot be accelerated is that the moisture has to migrate from the centre of the sweet to the surface. If an attempt is made to dry lozenges by microwave drying, the centre of the lozenge heats very rapidly as the water heats and flashes into steam. The steam then causes the lozenge to explode.

CREAM PASTE

Cream paste is a confectioner's name for these products. The name is not used commercially as it implies that the material is a dairy product which it is not. The commonest use of cream pastes is as the non-liquorice component of liquorice allsorts although cream paste products can vary from items as hard as lozenges through soft nut and fruit pastes to fruit noyaus (a confection made from icing sugar with a mucilage made from sugar, glucose syrup and a soft fruit). All of these products are made by mixing icing sugar with a mucilage which for a lozenge could be just gum tragacanth and water. In a fruit noyau the mucilage could be sugar, glucose syrup and raspberries; some caster sugar could be added to roughen the texture. However, a typical cream paste is made by mixing icing sugar with a mucilage made from glucose syrup, gelatine and a hard fat.

The Nature of Cream Paste

Inside the cream paste there is a mixed sugar syrup that is saturated with sucrose. This means that on cooling, the sucrose is likely to crystallise. Dispersed in this paste are the fat and any other insoluble ingredients.

Cream Paste Ingredients

Cream pastes are typically made from the following:

Sugar, which provides sweetness and bulk. The product depends upon the solubility limits of sucrose.

Glucose syrup, which provides bulk and resists drying out. It also controls the crystallisation of sucrose.

Invert syrup, which sweetens the product and lowers the water activity.

Hard fat, which gives body to the product and acts as a lubricant during extrusion.

Gelatine, which helps to emulsify the fat and holds the paste together during extrusion.

Starch, which is used to extend the product.

Coconut flour is added to a particular type of cream paste for liquorice allsorts to give the traditional texture and flavour.

Making Cream Paste

Cream paste is made by mixing the mucilage into the icing sugar with a heavy duty mixer such as a 'z' blade mixer. The mucilage can be made either by boiling or non-boiling methods. For a boiled mucilage some of the sugar, the glucose syrup and the invert syrup are heated together, the fat and gelatine are mixed in and the mucilage is then mixed into the icing sugar. Alternatively, a non-boiled mucilage is made by heating the glucose syrup, the invert syrup and the fat to 93 °C and then adding the gelatine. The mucilage is then mixed in as before. The non-boiled method can be used on a continuous basis by adding liquid fat to a hot sugar and glucose syrup. The gelatine solution is then added.

Shaping the Cream Paste

The finished cream paste can either be rolled into sheets or extruded. If the paste is to be layered the bottom layer is extruded onto a cooled, stainless steel band and the other colours of paste are then added one by one. The finished product is cooled and cut as required. The importance of the cooling is that the paste stiffens up as it cools, not only because the sugar syrups become more viscous, but also because the gelatine tends to gel.

Chapter 15

Tabletting

Tabletting is the area where sugar confectionery and pharmaceutical manufacture come closest, as the machines are similar the process can also be similar. Confectioners normally try to make tablets as hard as possible whereas pharmaceutical tablets are normally made to be fairly soft.

Tablets are made by compressing a powder in a die. Under pressure the powder fuses and is ejected as a tablet. The problem of tabletting is to produce a powder that will flow freely when poured into the die but bonds satisfactorily when compressed. Some materials are available in a directly compressible form. One example is sorbitol, which goes some way to explain the popularity of pressed sorbitol tablets with manufacturers. If directly compressible materials are not being used then the material must be granulated before being tabletted.

GRANULATION

Tabletting materials can be prepared by a number of different methods as outlined below.

Wet Granulation

Wet granulation can be used with most materials. However, it is not suitable for use with effervescent tablets since wetting the citric acid and sodium bicarbonate that they contain will start the reaction. In wet granulation a dough is made (just as in lozenge making). The powders to be granulated are screened followed by mixing with a granulating solution (a binder dissolved in water) to make a firm dough. Common binders are gum arabic, gelatine, modified starch or alginates. Thorough mixing of the correct amount of granulating solution is essential for a good quality product – this can take up to one hour. If too much granulating solution is used very hard granules are produced, whereas if insufficient granulating solution is used the granules will not bind properly. The dough can be tested by compressing it by hand to form a ball: a correctly made dough neither crumbles nor sticks when the dough ball is broken apart. The dough is then milled roughly to produce granules, and these granules are then dried on trays at around 50–60 °C for up to 24 hours. More sophisticated methods such as rotary driers, microwaves and fluidised beds can also be used to dry the granules. These can save a great deal of time. For example, using a fluidised bed drier, a batch of granules can be dried in just 20–30 minutes compared with the potential 24 hours for tray drying. All of this is in contrast with the situation with lozenges where accelerating the drying damages the product. The important difference, however, is that the granules are not the finished product; hence any distortion in their shape does not matter.

Fluidised Bed Granulation

An alternative to drying granules in a fluidised bed is to carry out the entire process in a fluidised bed granulator. In this system the material to be granulated is fluidised by an air or gas stream. The solids are thus forced to behave as a fluid. The binder solution is sprayed on to the powder which causes it to agglomerate into granules. The granules are dried in the air stream which can be heated to 40–80 °C. Other ingredients such as colour and flavour can also be mixed into the fluidised granules by the air stream.

This sort of system can prepare a batch of granules in 60–90 minutes. However, whereas fluidised bed granulation is rapid and flexible it is not without its problems. The presence of finely divided powders means that the apparatus itself is a potential explosion hazard, although, obviously, these systems are constructed and used in such a way as to be safe. One possibility is to use nitrogen instead of air thus providing an inert atmosphere. A further problem is that a fluidised bed granulator is not the easiest piece of equipment to set up, particularly compared with a wet granulation system.

Slugging

Neither of the previous methods is suitable for a water-sensitive material such as that used in an effervescent tablet. A method that is suitable is 'slugging'. Here the materials are compressed in a large die (2–5 cm diameter). The rough tablets produced are then ground down and re-compressed to make the finished product. The resulting tablets are less hard than those produced by wet granulation but this is not too surprising since there is no binder medium. The method obviously requires that the materials will bind to some extent, and while slugging requires relatively little equipment there is considerable wear on the dies, particularly in the initial stage.

Other Ways of Granulating

Several other methods which have been developed can be used to make granules. They include the use of pressure rollers as an alternative to slugging, using a rotating pan as an agglomerator, and extrusion of the granules.

AVOIDING GRANULATION

If the bulk ingredient is suitable then the product can be tabletted directly. A directly compressible form of sucrose is available but is relatively expensive. Fructose and dextrose can also be directly compressed. Specially made, directly compressible forms of the sugar substitutes sorbitol, mannitol and maltitol are available; a hygroscopic material like sorbitol would not be easy to wet granulate.

TABLETTING ADDITIVES

There are three types of lubricants added to the powders in order to deal with possible tabletting problems:

(1) *Glidants.* A poorly-flowing material will be treated by adding a glidant. This is particularly likely to be the case where the bulk ingredient is in a directly compressible form. The aim of a glidant is to reduce inter-particular friction, a typical glidant being a silica material that forms a coating around the particles of the base material. To do this the glidant should have a large surface area, *i.e.* it will have a small particle size. Excess levels of glidant have a detrimental effect upon the flow properties.

(2) *Pure lubricants.* Oils and fats can be used as lubricants although stearic acid or magnesium stearate are more common. The lubricant forms a hydrophobic surface between the die and the tablet, which allows the tablet to be ejected more easily from the die. Used to excess a lubricant will reduce the binding of the tablet, and used at too low a level the tablet tends to bind in the die. The lubricant can be added at the granulation stage or just prior to granulation, although adding at the granulation stage gives more effective mixing.

(3) *Anti-adherents.* These prevent the tablet adhering to the upper die. This is usually a problem where the tablets being made have complicated shapes.

Chapter 16

Experiments

These experiments have been devised on the basis that they should not use any ingredients that are not freely available in the UK. This means that recipes based on confectioner's glucose have been excluded.

HEALTH AND SAFETY

Two of these experiments involve boiling sugar solutions, and because of the concentration of sugar the boiling point is elevated above that of water. Sugary materials are sticky. It follows, therefore, that if a splash falls on anyone they will be hurt. It is incumbent on anyone carrying out these experiments to take precautions against contact with the hot syrup and to take precautions to deal with any accidental splashes. Treatment for any splash is immediate immersion in cold water or holding the afflicted part under cold, running water for several minutes. These experiments are not intended to be carried out by unsupervised children. If the experiment is conducted in a laboratory then the sugar will cease to be a food and must not be consumed.

SUGAR CRYSTALLISATION EXPERIMENTS

Use the method below and see the largest crystals that can be produced. Try dissolving and re-crystallising the sugar. Does this give bigger crystals? If sugar of higher purity than ordinary granulated can be obtained, try repeating the experiment.

Making the Syrup

The measuring vessel does not have to be a cup; any suitable vessel, *e.g.* a

151

beaker, will do. If more is wanted use a bigger vessel, if less, use a smaller one.

Dissolve $2\frac{1}{2}$ measures of sugar in 1 measure of water.
Cook without stirring to 118–112 °C (242–252 °F)

Small-scale Experiment

Make the syrup on a small scale. The syrup produced will be super-saturated when it cools. Pour some syrup into a test tube with a weighted string. After the syrup cools crystals will appear.

Questions:

(1) Which factors affect the rate of crystallisation?
(2) How can bigger crystals be produced?
(3) Is rapid crystallisation compatible with big crystals?
(4) What effect does using brown sugar have?
(5) Is the result improved by using a seed crystal?
(6) If purer sugar was wanted how would you obtain it?

Larger-scale Experiment

(1) Use an aluminium foil pie dish about 20 cm (8 inch) square or a similar vessel.
(2) Punch holes at the top edge of the dish.
(3) String about seven strings from one side to the other.
(4) Place the laced dish in a deeper pan to catch excess syrup.
(5) Pour the syrup into the dish until it is about 2 cm ($\frac{3}{4}$ inch) above the strings.
(6) Cover the surface with a piece of aluminium foil.
(7) Watch and wait. It can take a week to crystallise.
(8) Lift out the laced dish.
(9) Cut the strings and dislodge the crystals.
(10) Rinse quickly in cold water and place on a rack in a very low oven to dry.

Using your conclusions from the small-scale experiment how would you modify this experiment to produce a crop of very large crystals?

LOZENGE MAKING

Background

Lozenges are one of the oldest types of sugar confectionery. They are made by making a dough from icing sugar and a binding agent. The dough is then rolled out, cut to shape and the pieces are dried.

Try making up the recipe and taste the lozenges.

Method

Quantities:

- 300 g icing sugar;
- 28 g water;
  N.B. the water can equally be measured as a volume of 28 ml.
- 5 g gelatine;
- Peppermint flavour to test (other flavours can be used if desired).

(1) Dissolve the gelatine in the water.
(2) Mix the solution into the icing sugar to produce a dough.
(3) Mix the flavour into the dough to taste – as lozenges lose flavour upon drying it is best to be generous with the flavour.
(4) Roll out the dough and cut to shape. Any pieces left over can be re-rolled and cut.
(5) Dry the shaped pieces either in a low oven or in the air until they are hard.

Exercises

(1) Try making the product without using the gelatine.
(2) Other binding agents are commonly used to make lozenges, particularly gum tragacanth. Commercially, a mixture of gum tragacanth and gum acacia is used. Try substituting gum traga-canth for gelatine. (Remember that gum tragacanth needs to be allowed to swell and form a mucilage before use.)
(3) Try using gum acacia as a substitute for gelatine.
(4) Try a mixture of gum acacia and gum tragacanth, as in commer-cially produced lozenges.
(5) Compare the different products.
(6) Calculate the percentage difference in measuring the water by weight or volume at 25 °C. Is it likely to be significant?

N.B. any gum preparations used for making lozenge samples should be food grade and therefore fit for this use.

FUDGE (OR GRAINED CARAMEL)

Strictly speaking, this product is a grained caramel rather than a fudge; however, the end product will pass for a fudge.

- Part 1   2 tablespoons golden syrup
           3 tablespoons water
           84 g (3 oz) butter or margarine
           454 g (1 lb) caster sugar
- Part 2   1 small (397 g) can condensed milk
- Part 3   a few drops of vanilla essence

Method

(1) Put the ingredients for Part 1 into a heavy-based saucepan.
(2) Heat slowly until the sugar has dissolved.
(3) Add the condensed milk of Part 2.
(4) Bring to the boil.
(5) Boil until a sugar thermometer reads 116 °C (241 °F).
(6) Remove the pan from the heat and beat the mixture until it becomes thick.*
(7) Add Part 3 (the vanilla essence).
(8) Pour the product into a lightly greased tin, approximately 20 × 30 cm (8 × 12 inches) or equivalent.
(9) Mark the fudge into pieces with a knife when half set.
(10) When the product is cold, remove it from the tin.

Variations on the Recipe

It is possible to perform a number of experiments to discover the effects of variation in either the method or the ingredients – this covers any variability in the standard ingredients or the way in which the experiment was carried out. In carrying out these experiments it is always wise to make a control sample to the standard recipe. When experimenting it also pays to make sure that all the ingredients that are meant to be the same come from the same batch.

Method Variations

(1) Make a control batch and another batch where the product is not beaten on cooling but is poured directly into the tin. When the two

---

* The fudge can be beaten by hand or with a mixer. If a mixer is used it must be a robust table mixer like a Kenwood Chef. A hand mixer will have smoke pouring from it.

batches are cool examine them both with the naked eye and a microscope if available. Taste the products and comment. Continue to examine and taste the products over a number of days. Record your observations.

(2) Make the product, but starting with the condensed milk and adding the fat last. Compare the product with the control.

(3) Make batches using different fat sources, comparing butter and margarine or hard vegetable fat.

(4) Evaluate a similar batch using skimmed sweetened condensed milk but adding back all the milk fat as butter. Compare with the control.

(5) Make two batches, both using skimmed sweetened condensed milk, one with all the fat as butter, the other one with all margarine or vegetable fat. Compare.

(6) Evaluate a recipe based on skimmed milk powder but replacing all the milk fat with butter, whilst also allowing for the moisture content of the butter. Make up this recipe and compare it with a control recipe made from condensed milk. Do you now know why toffee is normally made from condensed milk?

(7) Repeat 6 but make the skimmed milk powder into a 'condensed milk substitute' with sugar and water and leave to stand in a cool place for 24 hours before use. Is the product better?

(8) Obtain some soft brown sugar. Make a control sample using full cream condensed milk and 100% white sugar, and make two batches, one with a 50% replacement, white for brown, of the added sugar, the other with 100% replacement. Are they distinguishable? Which do you prefer?

Taste Tests

Up to this point we have assumed that the product will be tasted by tasters who know which product is which. Another approach is to taste 'blind' where the taster does not know which piece comes from which batch. The aim of these tests is to find if there is a significant difference between the samples.

Now significant has two meanings: one ordinary and one statistical. The statistical meaning is related to confidence levels, and a type of statistical test that is used is the triangular test. Here, tasters are given three examples of the product under test: two are the same, one is not. Tasters are asked to identify which two samples are the same and which is different. There is always a chance that tasters will guess correctly, purely by chance. This probability can be calculated and is compared with the real result. The size of the difference can be related to the chance that the result obtained was done so by chance or represents a statistically significant difference.

This type of test depends upon the testers being representative of the group in question, *e.g.* fudge purchasers. If the testers are recruited from the staff of a confectionery factory there is a risk that they will become more discriminating than the general public.

Exercise

Consult a text book on statistics and set up a triangular test using two different variants of one of the recipes. Carry out the test and analyse the results. Repeat the exercise using two different batches of the same product which are identical except for the random variations that normally occur in cooking. Are they significantly different statistically?

Chapter 17

The Future

It is very difficult to predict the future of sugar confectionery. New ingredients are likely to become available and this should logically lead to new products. It might, however, lead simply to new versions of old products. It appears a safe prediction that the Rowntree fruit pastille will continue to be sold in the 21st century as they have been for part of the 19th century and throughout the 20th century. The £4 a week that Claud August Gaget was paid in the 1880s while he invented them has to be one of the best investments that any company has ever made in research and development. The fundamental truth that finding a genius and paying them well is a good investment has probably been lost under layers of corporate culture.

One possibility for the future is the growth of sugar-free products. This is likely to need the problems of existing sugar-free confectionery to be overcome – the laxative threshold is, after all, a very effective limit on consumption. One sign that, if the problems of sugar-free product were overcome, it would become the dominant product is the position in chewing gum where sugar-free outsells sugar-containing. Of course, some individuals would not benefit from the energy reductions in sugar-free products any way. The success of tooth friendly products remains to be seen.

Subject Index